Student Edition

Eureka Math
Grade K
Module 1

Special thanks go to the Gordon A. Cain Center and to the Department of
Mathematics at Louisiana State University for their support in the development of
Eureka Math.

For a free *Eureka Math* Teacher
Resource Pack, Parent Tip
Sheets, and more please
visit www.Eureka.tools

Published by the non-profit Great Minds

Copyright © 2016 Great Minds. No part of this work may be reproduced, sold, or commercialized, in whole or in part, without written permission from Great Minds. Non-commercial use is licensed pursuant to a Creative Commons Attribution-NonCommercial-ShareAlike 4.0 license; for more information, go to http://greatminds.net/maps/math/copyright. "Great Minds" and "Eureka Math" are registered trademarks of Great Minds.

Printed in the U.S.A.

This book may be purchased from the publisher at eureka-math.org

10 9 8 7

ISBN 978-1-63255-873-2

Name _____   Date _____

Find animals that are exactly the same. Then, find animals that look like each other but are not exactly the same. Use a ruler to draw a line connecting the animals.

Lesson 1:   Analyze to find two objects that are *exactly the same* or *not exactly the same*.

©2016 Great Minds. eureka-math.org
GK-M1-SE-B1-1.3.1-01.2016

1

This page intentionally left blank

Name _____   Date _____

Color the things that are exactly the same. Color them so that they look like each other.

Lesson 1:   Analyze to find two objects that are *exactly the same* or *not exactly the same*.

3

This page intentionally left blank

4

Name _____ Date _____

Use your ruler to draw a line between two objects that match.
Use your words. "These are the same, but this one _____, and this one _____."

EUREKA MATH™

©2016 Great Minds. eureka-math.org
GK-M1-SE-B1-1.3.1-01.2016

This page intentionally left blank

Name _____   Date _____

Draw a line between two objects that match. Use your words. "These are the same, but this one _____, and this one _____."

©2016 Great Minds. eureka-math.org
GK-M1-SE-B1-1.3.1-01.2016

This page intentionally left blank

Name _____   Date _____

Draw a line between the objects that have the same pattern. Talk with a neighbor about the objects that match.

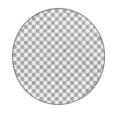

Circle the object that would be used together with the object on the left.

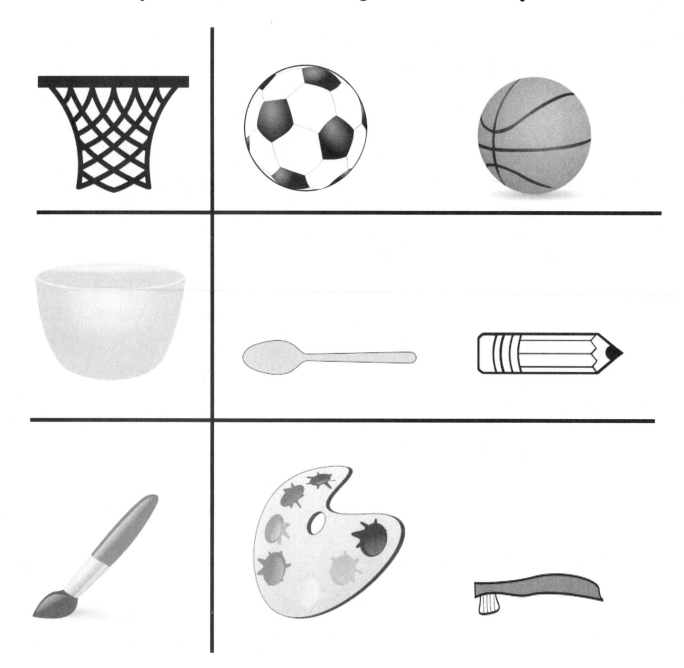

Lesson 3:   Classify to find two objects that share a visual pattern, color, and use.

Name _____   Date _____

Draw something that you would use with each. Tell why.

Make a picture of 2 things you use together. Tell why.

EUREKA
MATH™

Lesson 3:   Classify to find two objects that share a visual pattern, color, and use.

11

©2016 Great Minds. eureka-math.org
GK-M1-SE-B1-1.3.1-01.2016

This page intentionally left blank

Name _____   Date _____

Use the cutouts. Glue the pictures to show where to keep each thing.

This page intentionally left blank

Problem Set Cutouts

Lesson 4:   Classify items into two pre-determined categories.

15

©2016 Great Minds. eureka-math.org
GK-M1-SE-B1-1.3.1-01.2016

This page intentionally left blank

Name _____   Date _____

Make two groups. Circle things that belong to one group. Underline the things that belong to the other group. Tell someone why the items in each group belong together. (There is more than one way to make groups!)

This page intentionally left blank

Name _____   Date _____

Draw a line with your ruler to show where each thing belongs.

Lesson 5:   Classify items into three categories, determine the count in each, and
reason about how the last number named determines the total.

19

©2016 Great Minds. eureka-math.org
GK-M1-SE-B1-1.3.1-01.2016

This page intentionally left blank

Name _____   Date _____

Use the cutouts. Glue the pictures to show where each belongs. Tell an adult how many are in each place.

Library

School

Grocery Store

Lesson 5:   Classify items into three categories, determine the count in each, and reason about how the last number named determines the total.

21

©2016 Great Minds. eureka-math.org
GK-M1-SE-B1-1.3.1-01.2016

This page intentionally left blank

Homework Cutouts

Lesson 5:   Classify items into three categories, determine the count in each, and
reason about how the last number named determines the total.

23

This page intentionally left blank

Name _____   Date _____

Count and color.

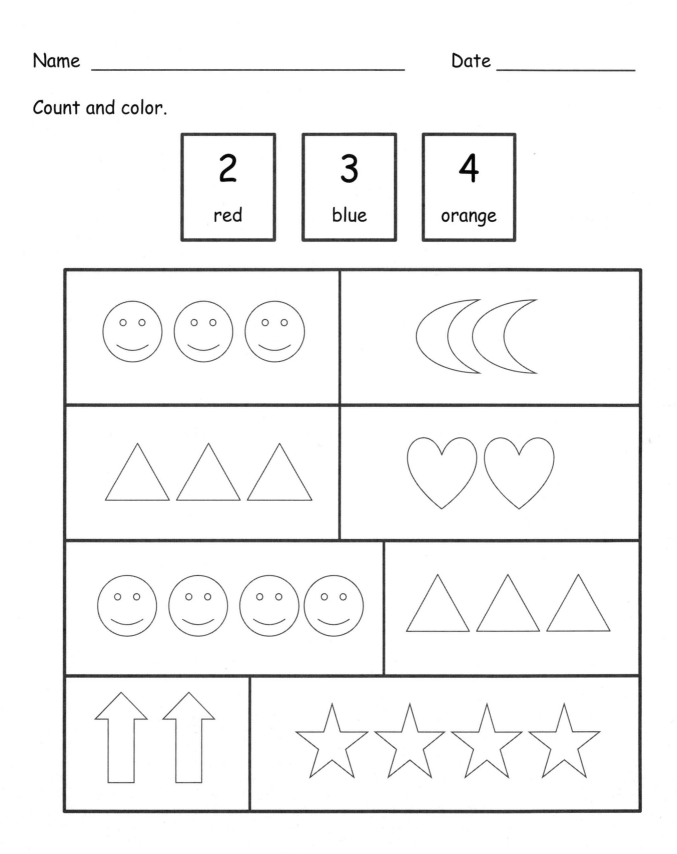

EUREKA MATH   **Lesson 6:**   Sort categories by count. Identify categories with 2, 3, and 4 within a given scenario.   25

©2016 Great Minds. eureka-math.org
GK-M1-SE-B1-1.3.1-01.2016

This page intentionally left blank

Name _____   Date _____

Draw lines to put the treasures in the boxes.

Lesson 6:   Sort categories by count. Identify categories with 2, 3, and 4 within a
given scenario.

27

©2016 Great Minds. eureka-math.org
GK-M1-SE-B1-1.3.1-01.2016

This page intentionally left blank

Name _____   Date _____

Count and color.

EUREKA MATH™

Lesson 7:   Sort by count in vertical columns and horizontal rows (linear configurations to 5). Match to numerals on cards.

29

This page intentionally left blank

Name _____   Date _____

Count and color.

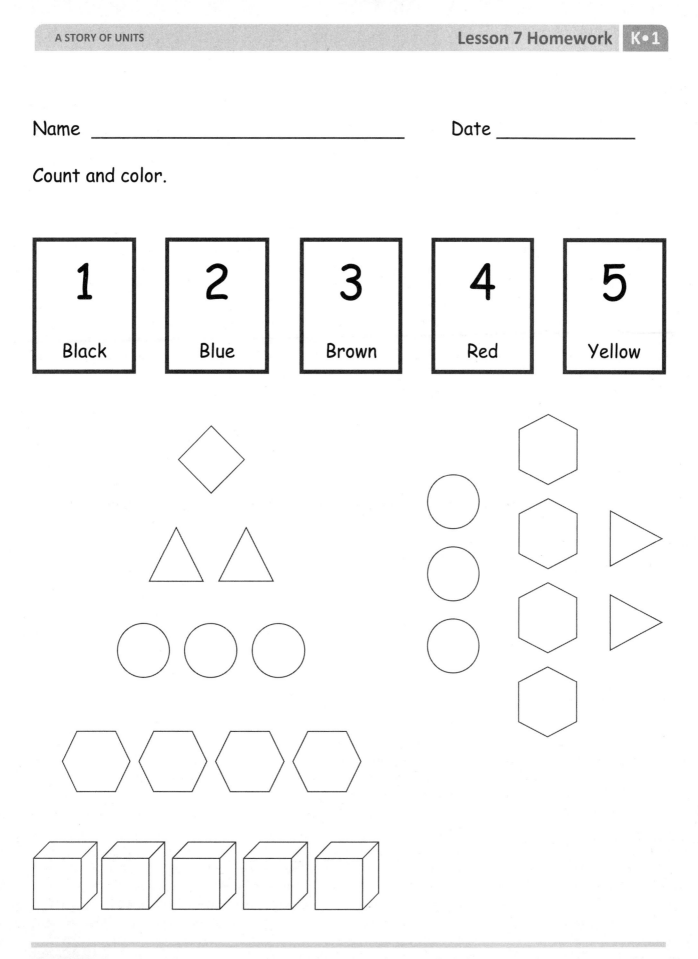

EUREKA
MATH

Lesson 7:   Sort by count in vertical columns and horizontal rows (linear configurations to 5). Match to numerals on cards.

31

©2016 Great Minds. eureka-math.org
GK-M1-SE-B1-1.3.1-01.2016

This page intentionally left blank

Name _____   Date _____

Count the objects. Circle the correct number.

1 2 3	1 2 3
3 4 5	2 3 4
4 3 2	5 4 1
4 3 2	5 4 1

EUREKA MATH™

Lesson 8:   Answer *how many* questions to 5 in linear configurations (5-group), with 4 in an array configuration. Compare ways to count five fingers.

33

©2016 Great Minds. eureka-math.org
GK-M1-SE-B1-1.3.1-01.2016

This page intentionally left blank

Name _____   Date _____

Count. Circle the number that tells how many dots in all.

● ● ● ●	4	5
• • • •	4	5
● ● ● ● ●	4	5
• • • • •	4	5
[die: 4]	4	5
[die: 4] [die: 1]	4	5
[die: 5]	4	5

Lesson 8:   Answer *how many* questions to 5 in linear configurations (5-group),
with 4 in an array configuration. Compare ways to count five fingers.

35

This page intentionally left blank

Name _____   Date _____

Count the dots, and circle the correct number. Color the same number of dots on the right as the gray ones on the left to show the hidden partners.

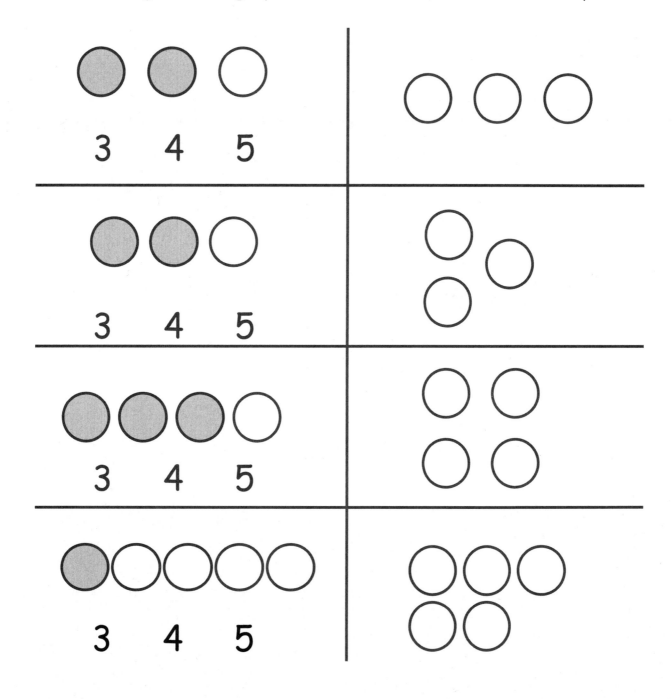

Lesson 9:   Within linear and array dot configurations of numbers 3, 4, and 5, find
            hidden partners.

37

This page intentionally left blank

Name _____   Date _____

Count the circles, and box the correct number. Color the same number of circles on the right as the shaded ones on the left to show hidden partners.

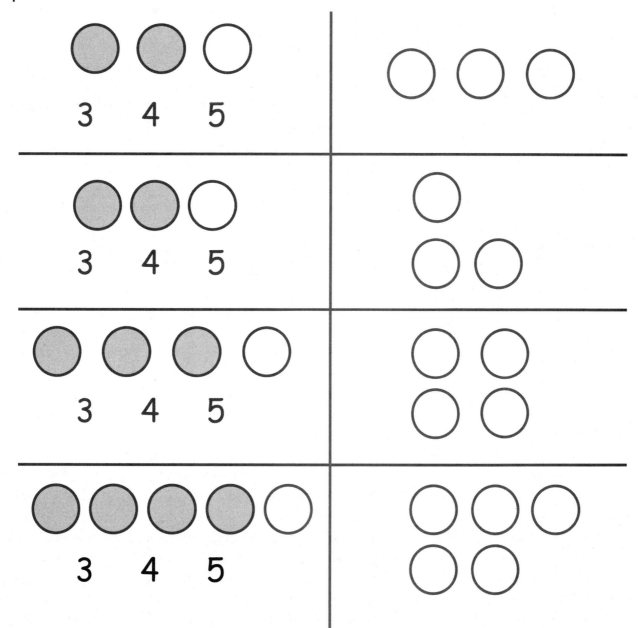

Lesson 9: Within linear and array dot configurations of numbers 3, 4, and 5, find *hidden partners*.

39

©2016 Great Minds. eureka-math.org
GK-M1-SE-B1-1.3.1-01.2016

This page intentionally left blank

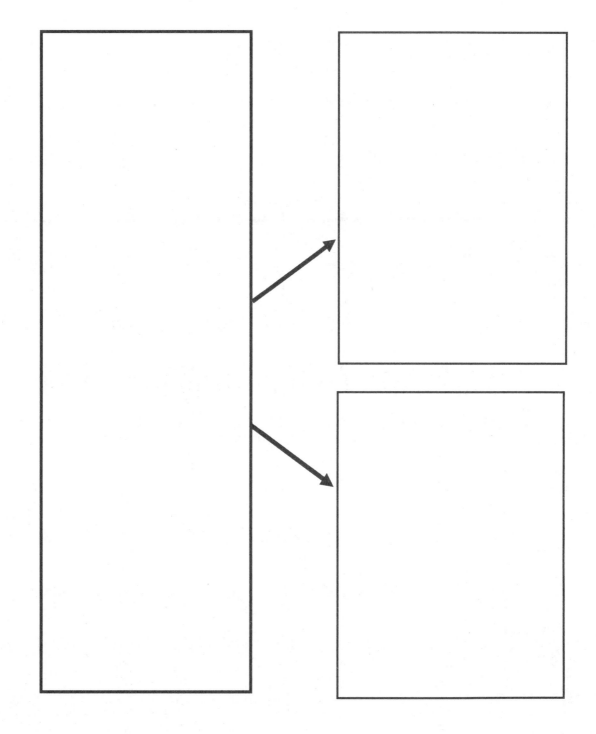

hidden partners

This page intentionally left blank

Name _____   Date _____

Color to see the hidden partners.

Count the objects. Circle the total number.

Color 1 circle.	Color 1 star.	Color 1 circle.
1 2 3	2 3 4	3 4 5
Color 2 stars.	Color 2 circles.	Color 2 stars.
3 2 1	5 4 3	4 5 3

Draw 2 circles. Count all the objects, and circle the number.

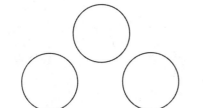

5   2   3

Lesson 10:   Within circular and scattered dot configurations of numbers 3, 4,
and 5, find *hidden partners*.

43

This page intentionally left blank

Name _____   Date _____

Count how many. Draw a box around that number. Then, color 1 of the circles in each group.

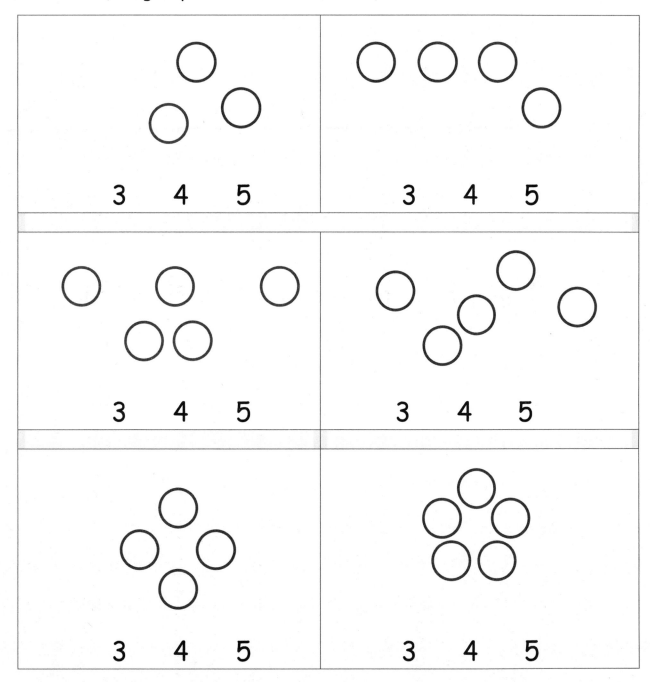

Talk to an adult at home about the hidden partners you found.

Lesson 10:   Within circular and scattered dot configurations of numbers 3, 4, and 5, find *hidden partners*.

45

©2016 Great Minds. eureka-math.org
GK-M1-SE-B1-1.3.1-01.2016

This page intentionally left blank

Name _____ Date _____

Count the squares. Draw the squares above the numbers.

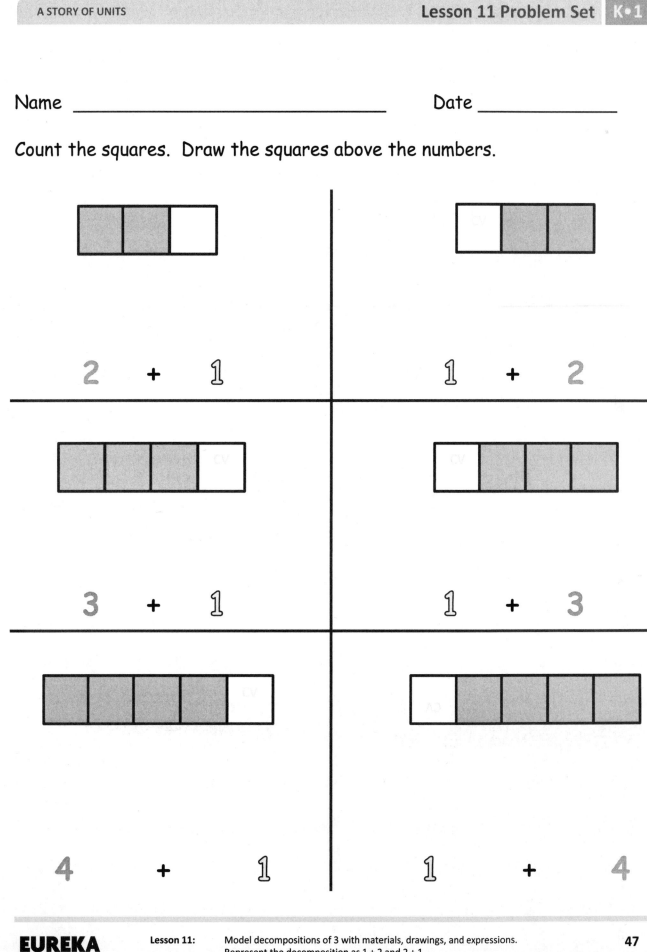

2   +   1

1   +   2

3   +   1

1   +   3

4   +   1

1   +   4

EUREKA MATH

Lesson 11:   Model decompositions of 3 with materials, drawings, and expressions.
Represent the decomposition as 1 + 2 and 2 + 1.

47

This page intentionally left blank

Name _____ Date _____

Feed the puppies! Here are 3 bones. Draw lines to show 2 + 1.

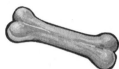 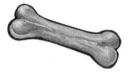 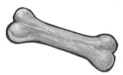

Color the shapes to show 1 + 4. Use your 2 favorite colors.

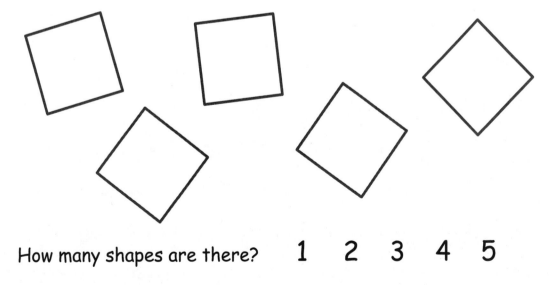

How many shapes are there?   1   2   3   4   5

Lesson 11:   Model decompositions of 3 with materials, drawings, and expressions.
Represent the decomposition as 1 + 2 and 2 + 1.

49

©2016 Great Minds. eureka-math.org
GK-M1-SE-B1-1.3.1-01.2016

This page intentionally left blank

Name _____   Date _____

Write 0.

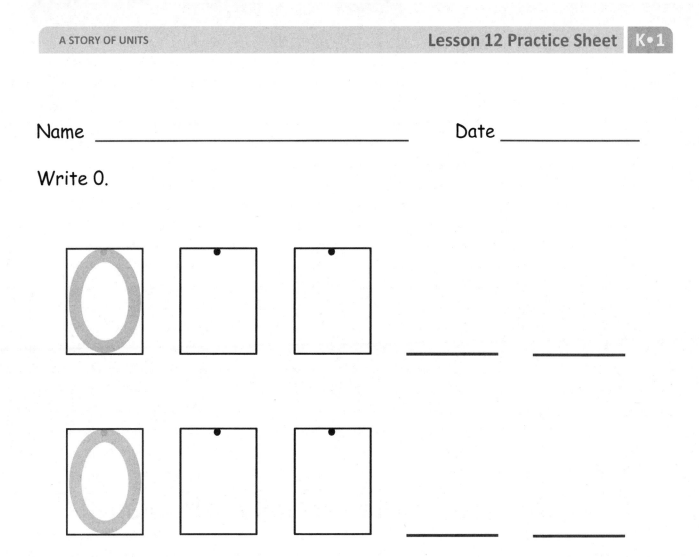

Lesson 12:   Understand the meaning of zero. Write the numeral 0.   51

This page intentionally left blank

Name _____   Date _____

Circle the number that tells how many.

0 1 2 3	0 1 2 3	0 1 2 3	0 1 2 3

0 1 2 3   0 1 2 3   0 1 2 3   0 1 2 3

How many elephants are in the trees?

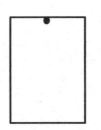

This page intentionally left blank

Name _____   Date _____

How many? Draw a line between each picture and its number.

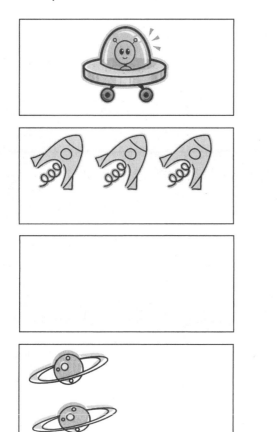

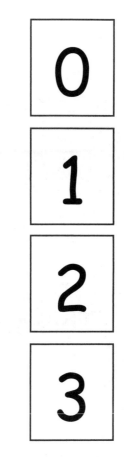

Write the number in the blank.

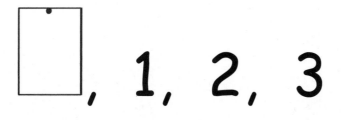

Lesson 12:   Understand the meaning of zero. Write the numeral 0.   55

©2016 Great Minds. eureka-math.org
GK-M1-SE-B1-1.3.1-01.2016

This page intentionally left blank

Name _____ Date _____

Write 1, 2, and 3.

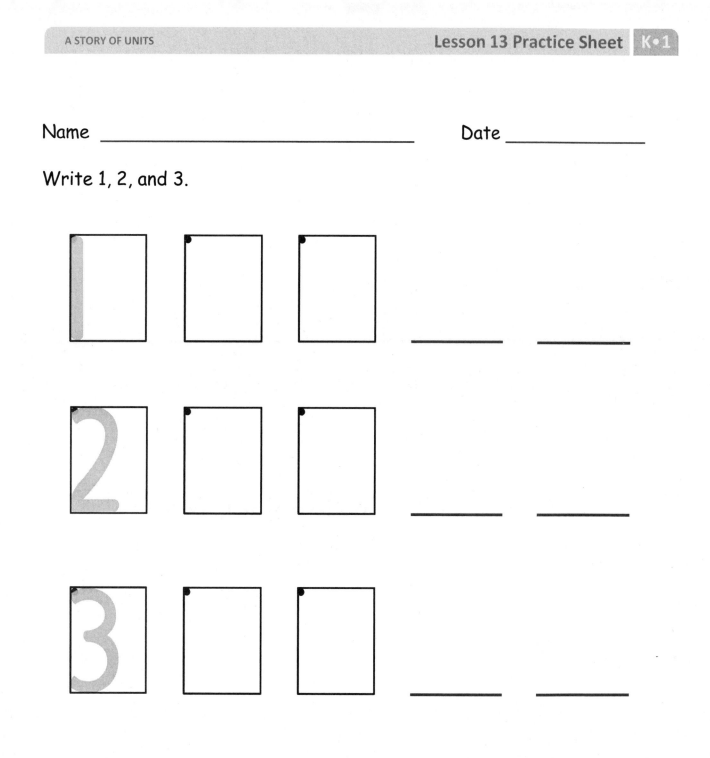

This page intentionally left blank

Name _____   Date _____

Write the missing numbers.

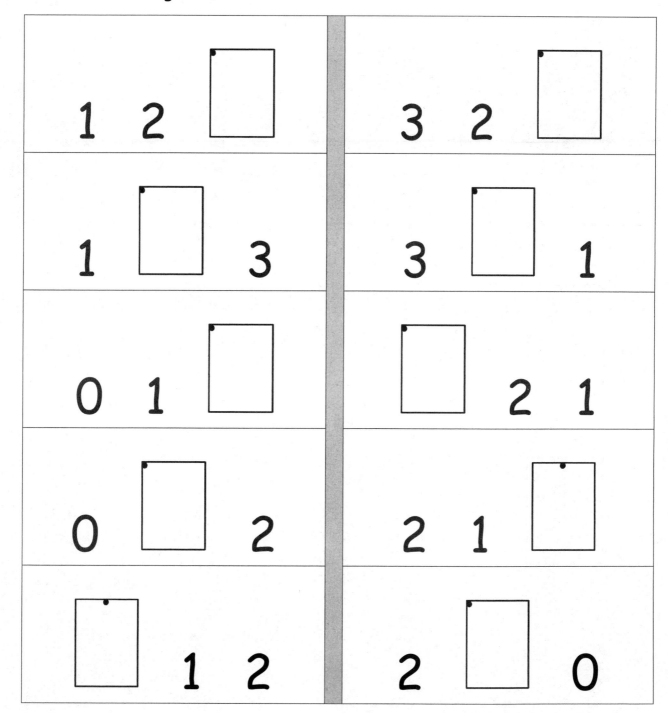

Count and write how many.

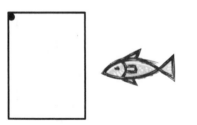

Lesson 13:   Order and write numerals 0–3 to answer *how many* questions.

©2016 Great Minds. eureka-math.org
GK-M1-SE-B1-1.3.1-01.2016

Name _____   Date _____

Draw ▣ (two) pots.

How many?

Draw ▣ (one) friend.

How many?

Draw ▣ (three) toys.

How many?

Lesson 13:   Order and write numerals 0–3 to answer *how many* questions.

61

©2016 Great Minds. eureka-math.org
GK-M1-SE-B1-1.3.1-01.2016

How many pet monkeys 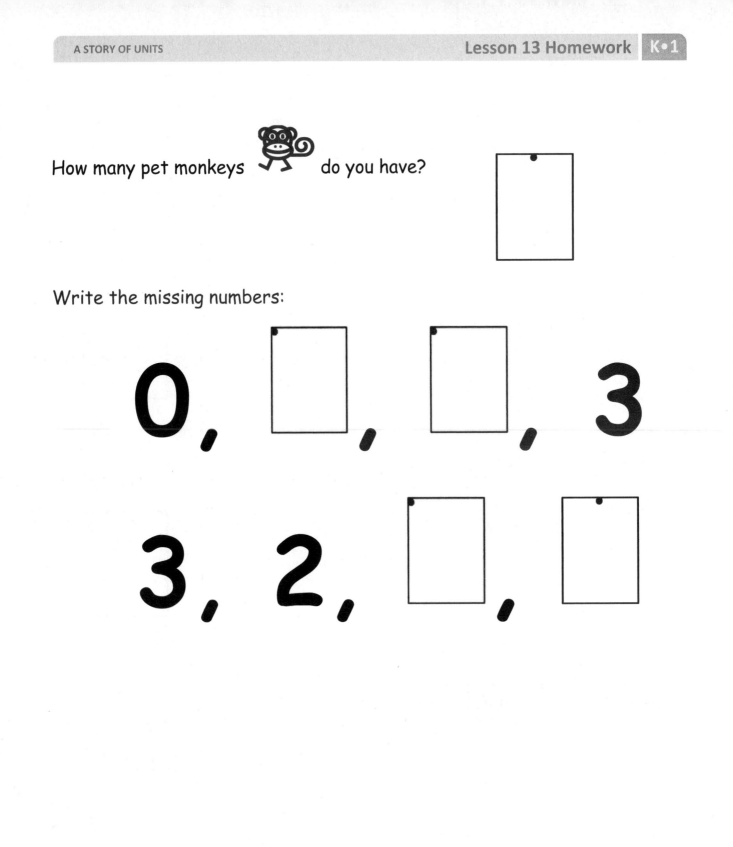 do you have?

Write the missing numbers:

0, ☐, ☐, 3

3, 2, ☐, ☐

Name _____   Date _____

Color the picture to match the number sentence.

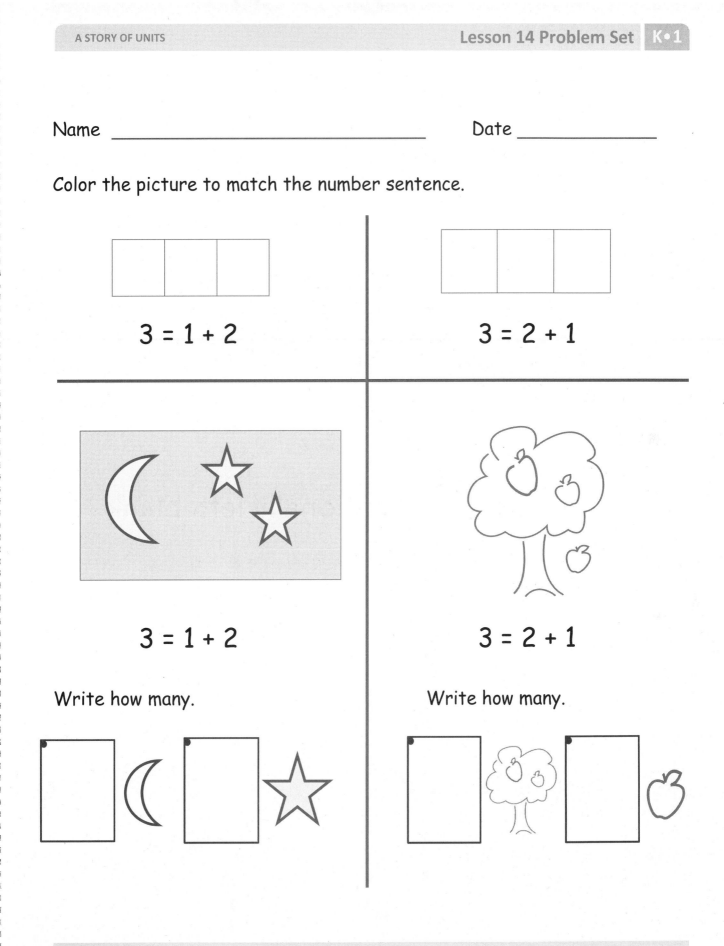

$3 = 1 + 2$

$3 = 2 + 1$

$3 = 1 + 2$

$3 = 2 + 1$

Write how many.

Write how many.

EUREKA MATH

Lesson 14:   Write numerals 1–3. Represent decompositions with materials,
drawings, and equations, 3 = 2 + 1 and 3 = 1 + 2.

63

©2016 Great Minds. eureka-math.org
GK-M1-SE-B1-1.3.1-01.2016

This page intentionally left blank

Name _____   Date _____

Color the shirts so that 1 is red and 2 are green.

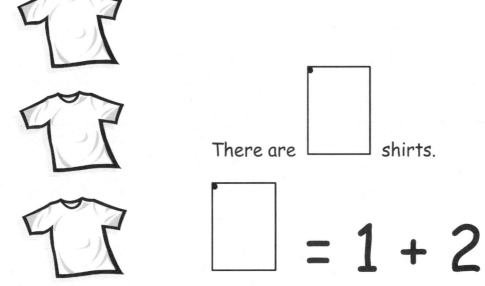

There are ☐ shirts.

☐ = 1 + 2

Color the balls so that 2 are yellow and 1 is blue.

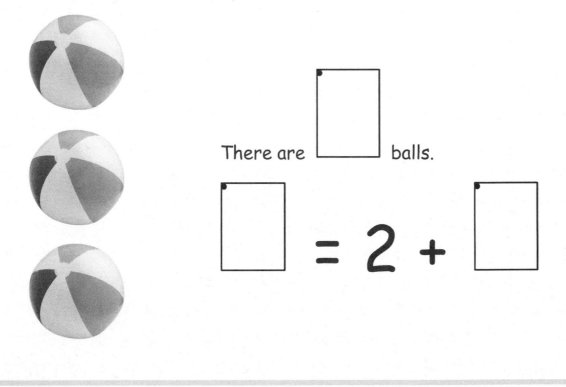

There are ☐ balls.

☐ = 2 + ☐

EUREKA
MATH™

Lesson 14:   Write numerals 1–3. Represent decompositions with materials,
drawings, and equations, 3 = 2 + 1 and 3 = 1 + 2.

65

This page intentionally left blank

Name _____   Date _____

Write 4 and 5.

Write the missing numbers:

1, 2, 3, ☐, ☐,

☐, ☐, 3, 2, 1

Lesson 15:   Order and write numerals 4 and 5 to answer *how many* questions in
categories; sort by count.

67

©2016 Great Minds. eureka-math.org
GK-M1-SE-B1-1.3.1-01.2016

This page intentionally left blank

Name _____ Date _____

Count and write how many. Circle a group of four of each fruit.

EUREKA MATH™

Lesson 15: Order and write numerals 4 and 5 to answer *how many* questions in categories; sort by count.

©2016 Great Minds. eureka-math.org
GK-M1-SE-B1-1.3.1-01.2016

This page intentionally left blank

Name _____   Date _____

Count the shapes and write the numbers. Mark each shape as you count.

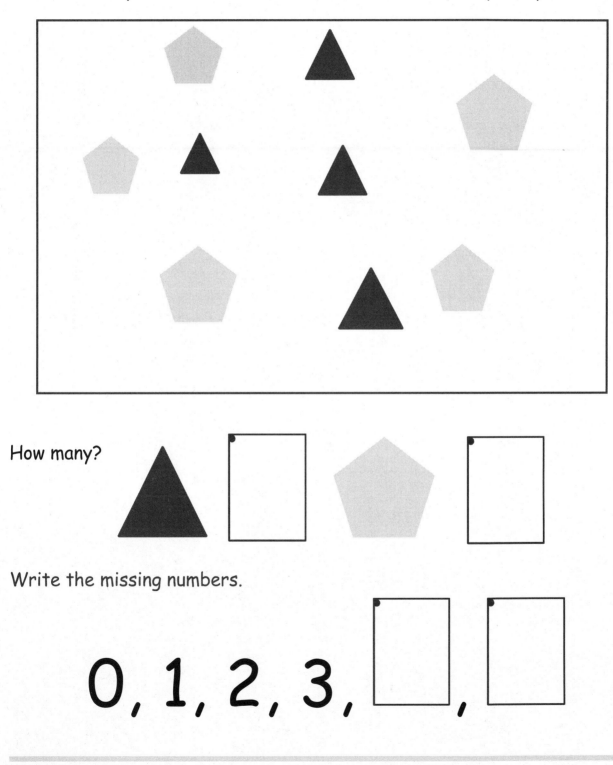

How many?

Write the missing numbers.

0, 1, 2, 3,

Lesson 15:   Order and write numerals 4 and 5 to answer *how many* questions in categories; sort by count.

©2016 Great Minds. eureka-math.org
GK-M1-SE-B1-1.3.1-01.2016

This page intentionally left blank

Name _____   Date _____

In each picture, color some squares red and some blue. Do it a different way each time.

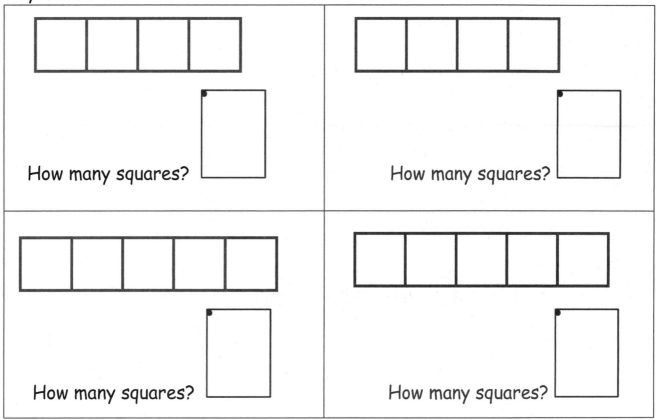

How many squares?

How many squares?

How many squares?

How many squares?

Draw more circles to make 4.

○ ○ ○	○ ○	○

Draw more X's to make 5.

XXXX	XXX	XX	X

Lesson 16:   Write numerals 1–5 in order. Answer and make drawings of decompositions with totals of 4 and 5 without equations.

73

©2016 Great Minds. eureka-math.org
GK-M1-SE-B1-1.3.1-01.2016

This page intentionally left blank

Name _____   Date _____

How many 🌙 ?

How many 🌙 ?

How many altogether?

How many 🙂 ?

How many 🙁 ?

How many altogether?

EUREKA
MATH™

Lesson 16:   Write numerals 1–5 in order. Answer and make drawings of
decompositions with totals of 4 and 5 without equations.

75

©2016 Great Minds. eureka-math.org
GK-M1-SE-B1-1.3.1-01.2016

This page intentionally left blank

Name _____ Date _____

Write the missing numbers:

1, 2, ☐, 4, ☐

5, ☐, 3, 2, ☐

☐, 3, 2, 1, ☐

☐, 1, 2, ☐, 4

Lesson 16:   Write numerals 1–5 in order. Answer and make drawings of decompositions with totals of 4 and 5 without equations.

77

Draw 3 red fish and 1 green fish.

How many fish are there in all?

3 fish and 1 fish make ⬜ fish.

Draw 2 happy faces and 3 sad faces.

How many faces are there in all?

5 is the same as ⬜ and ⬜.

Lesson 16: Write numerals 1–5 in order. Answer and make drawings of decompositions with totals of 4 and 5 without equations.

©2016 Great Minds. eureka-math.org
GK-M1-SE-B1-1.3.1-01.2016

Name _____ Date _____

Draw 1 more. Then count the objects, and write the number in the box.

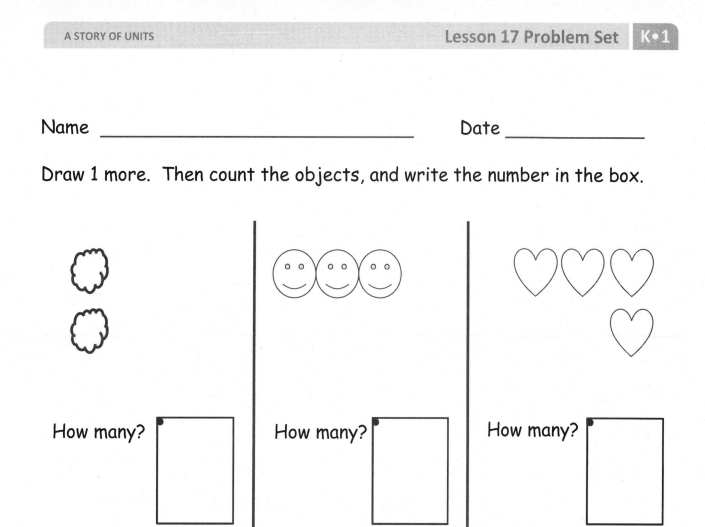

How many? ☐

How many? ☐

How many? ☐

Draw 1 more.
Then, circle the number.

Draw 6 fingers.

Draw 6 beads.

4   5   6

EUREKA
MATH™

Lesson 17: Count 4–6 objects in vertical and horizontal linear configurations and array configurations. Match 6 objects to the numeral 6.

79

©2016 Great Minds. eureka-math.org
GK-M1-SE-B1-1.3.1-01.2016

This page intentionally left blank

Name _____   Date _____

Color 4.

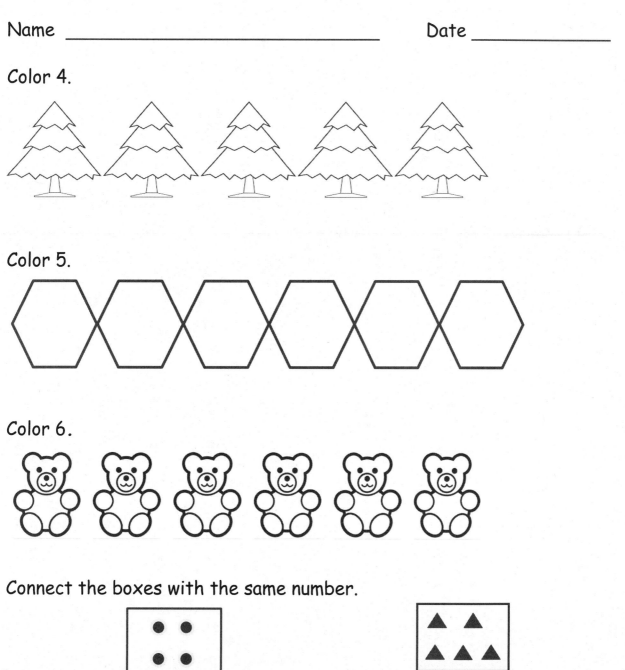

Color 5.

Color 6.

Connect the boxes with the same number.

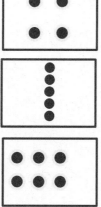

Lesson 17:   Count 4–6 objects in vertical and horizontal linear configurations and
array configurations. Match 6 objects to the numeral 6.

81

©2016 Great Minds. eureka-math.org
GK-M1-SE-B1-1.3.1-01.2016

This page intentionally left blank

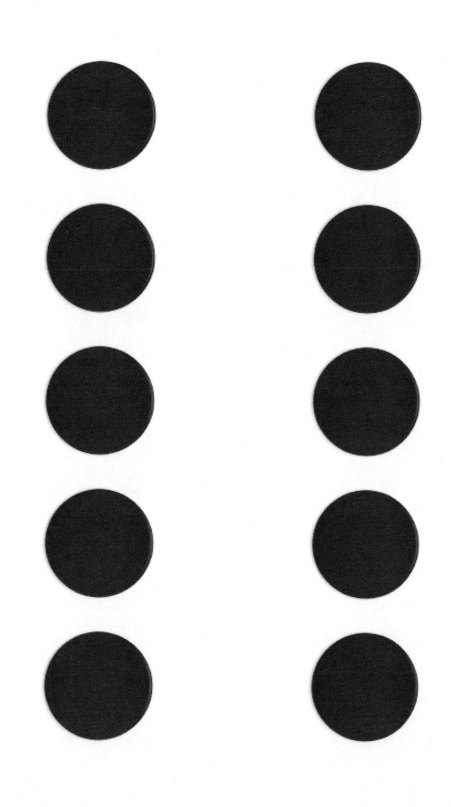

5-group mat

Lesson 17:   Count 4–6 objects in vertical and horizontal linear configurations and
array configurations. Match 6 objects to the numeral 6.

©2016 Great Minds. eureka-math.org
GK-M1-SE-B1-1.3.1-01.2016

83

This page intentionally left blank

Name _____   Date _____

Write 6.

Write the missing numbers:

$$1, 2, 3, 4, \boxed{}, \boxed{}$$

$$\boxed{}, \boxed{}, 4, 3, 2, 1$$

numeral formation practice sheet 6

Lesson 18:   Count 4–6 objects in circular and scattered configurations. Count 6 items out of a larger set. Write numerals 1–6 in order.

85

©2016 Great Minds. eureka-math.org
GK-M1-SE-B1-1.3.1-01.2016

This page intentionally left blank

Name _____   Date _____

Color 6 beans in each group.

Count the dots in each box. Write the number.

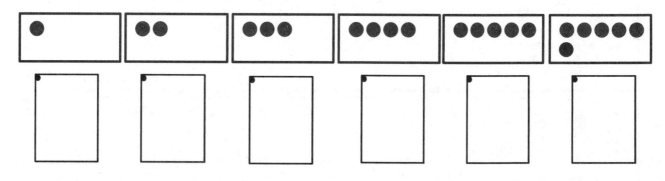

Lesson 18:   Count 4–6 objects in circular and scattered configurations. Count 6
              items out of a larger set. Write numerals 1–6 in order.

87

©2016 Great Minds. eureka-math.org
GK-M1-SE-B1-1.3.1-01.2016

Count the objects. Write the number in the box.

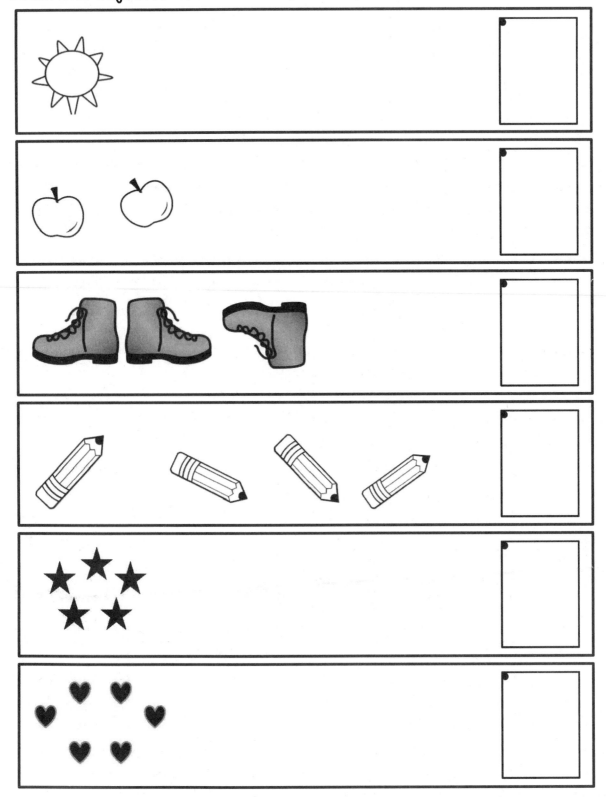

Lesson 18: Count 4–6 objects in circular and scattered configurations. Count 6 items out of a larger set. Write numerals 1–6 in order.

©2016 Great Minds. eureka-math.org
GK-M1-SE-B1-1.3.1-01.2016

Name _____   Date _____

Color 5.   Color 6.

Circle 6 🎈 balloons.

EUREKA MATH™

Lesson 18:   Count 4–6 objects in circular and scattered configurations. Count 6
items out of a larger set. Write numerals 1–6 in order.

©2016 Great Minds. eureka-math.org
GK-M1-SE-B1-1.3.1-01.2016

This page intentionally left blank

Name _____   Date _____

Color 5 in each group.

Color 5. Draw 2 circles to the right. Count all the circles.

Color 5. Draw 2 circles below. Count all the circles.

EUREKA
MATH™

Lesson 19:   Count 5–7 linking cubes in linear configurations. Match with numeral 7. Count on fingers from 1 to 7 and connect to 5 group images.

91

©2016 Great Minds. eureka-math.org
GK-M1-SE-B1-1.3.1-01.2016

This page intentionally left blank

Name _____   Date _____

Draw a line from the 5-groups to the matching numbers.

⬤⬤⬤☐☐	☐☐☐☐☐	**3**
⬤⬤⬤⬤⬤	⬤☐☐☐☐	**4**
⬤⬤⬤⬤☐	☐☐☐☐☐	**5**
⬤⬤⬤⬤⬤	☐☐☐☐☐	**6**
⬤⬤⬤⬤⬤	⬤⬤☐☐☐	**7**

Fill in the missing numbers.

1, ☐, 3, ☐, 5, ☐, 7

3, ☐, 5, ☐, 7

Lesson 19:   Count 5–7 linking cubes in linear configurations. Match with numeral
7. Count on fingers from 1 to 7 and connect to 5 group images.   93

©2016 Great Minds. eureka-math.org
GK-M1-SE-B1-1.3.1-01.2016

This page intentionally left blank

Name _____   Date _____

Write 7.

Write the missing numbers:

___ , 2, 3, 4, 5, ___ , ___

7, 6, ___ , 4, 3, ___ , ___

numeral formation practice sheet 7

Lesson 20: Reason about sets of 7 varied objects in circular and scattered configurations. Find a path through the scattered configuration. Write numeral 7. Ask, "How is your seven different from mine?"

95

©2016 Great Minds. eureka-math.org
GK-M1-SE-B1-1.3.1-01.2016

This page intentionally left blank

Name _____   Date _____

Color 7 beans. Draw a line to connect the beans you colored.

Color 7 beans.

Count the dots in each box. Write the number.

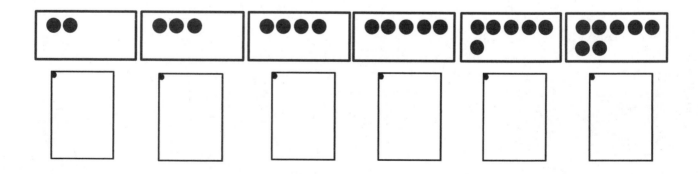

 EUREKA MATH™

Lesson 20:   Reason about sets of 7 varied objects in circular and scattered
configurations. Find a path through the scattered configuration.
Write numeral 7. Ask, "How is your seven different from mine?"

97

©2016 Great Minds. eureka-math.org
GK-M1-SE-B1-1.3.1-01.2016

This page intentionally left blank

Name _____   Date _____

How many? Write the number in the box.

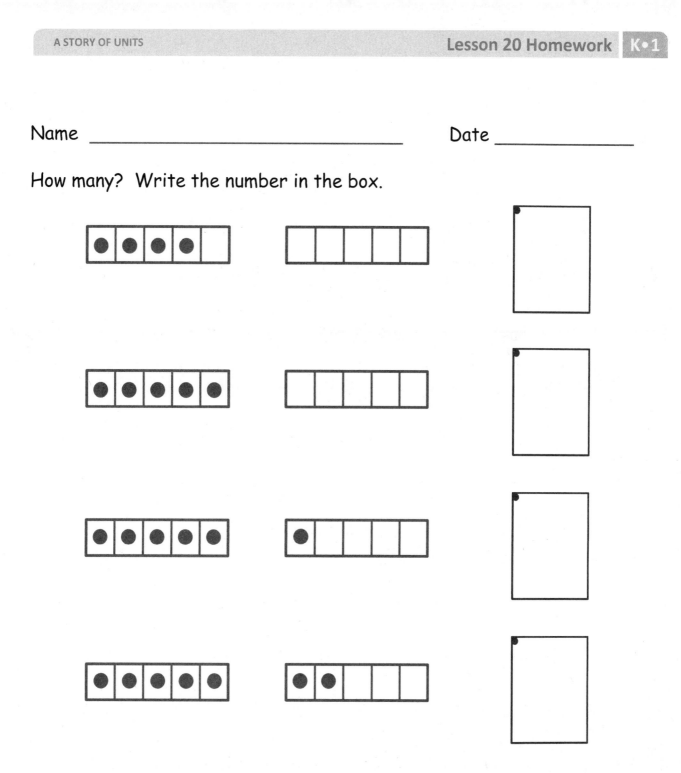

Lesson 20:   Reason about sets of 7 varied objects in circular and scattered
configurations. Find a path through the scattered configuration.
Write numeral 7. Ask, "How is your seven different from mine?"

99

EUREKA
MATH™

Count how many. Write the number in the box.
Draw a line to show how you counted the suns.

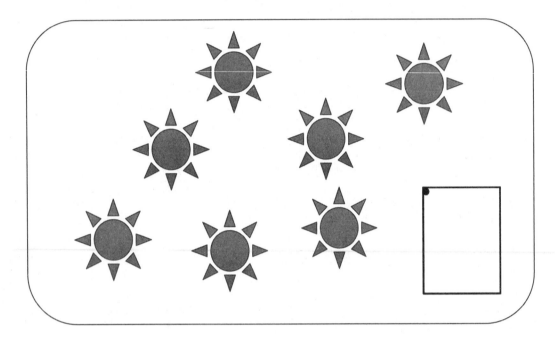

Count how many. Write the number in the box.
Draw a line to show how you counted the circles.

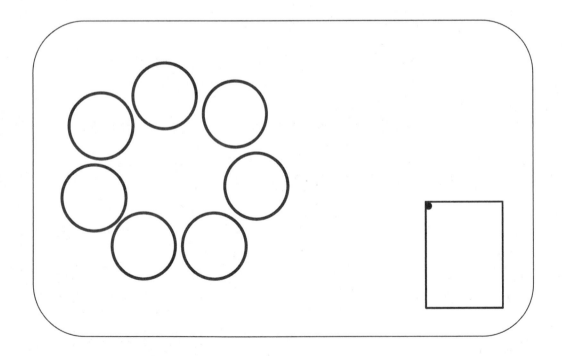

Lesson 20: Reason about sets of 7 varied objects in circular and scattered
configurations. Find a path through the scattered configuration.
Write numeral 7. Ask, "How is your seven different from mine?"

©2016 Great Minds. eureka-math.org
GK-M1-SE-B1-1.3.1-01.2016

Name _____   Date _____

Count and circle how many. Color 5.

6   7   8

6   7   8

| Color 5 circles. Draw 3 circles to the right. Count all the circles. | Color 5 circles. Draw 3 circles below. Count all the circles. |

  EUREKA MATH™   Lesson 21:   Compare counts of 8 in linear and array configurations. Match with numeral 8.   101

©2016 Great Minds. eureka-math.org
GK-M1-SE-B1-1.3.1-01.2016

Count and circle how many. Color 4.

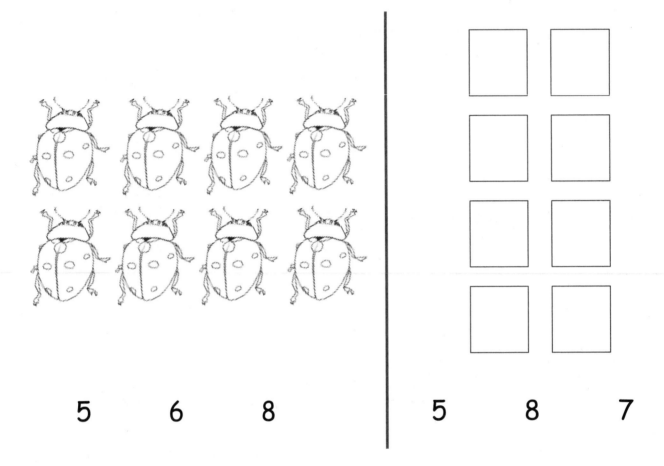

5   6   8   5   8   7

Color 4. Then, draw 3 circles to finish the row.

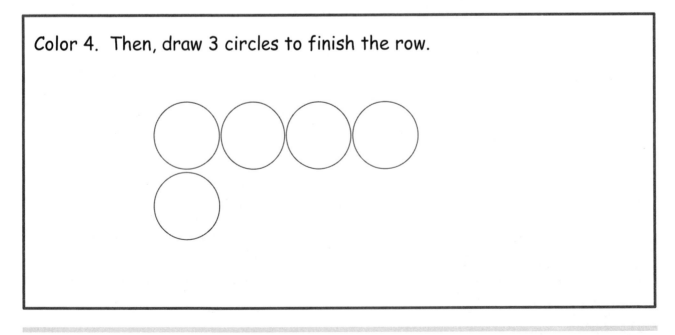

Lesson 21: Compare counts of 8 in linear and array configurations. Match with numeral 8.

Name _____ Date _____

Color 4 squares blue. Color 4 squares yellow.

Count and circle how many.

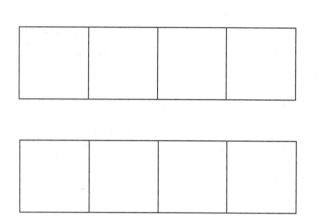

6   7   8

Color 4 squares green. Color 4 squares brown.

Count and circle how many.

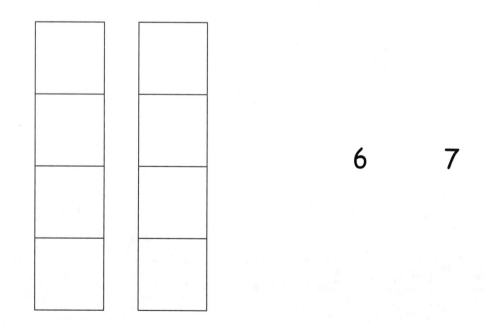

6   7   8

Lesson 21:   Compare counts of 8 in linear and array configurations. Match with
numeral 8.

103

©2016 Great Minds. eureka-math.org
GK-M1-SE-B1-1.3.1-01.2016

Count how many. Write the number in the box.

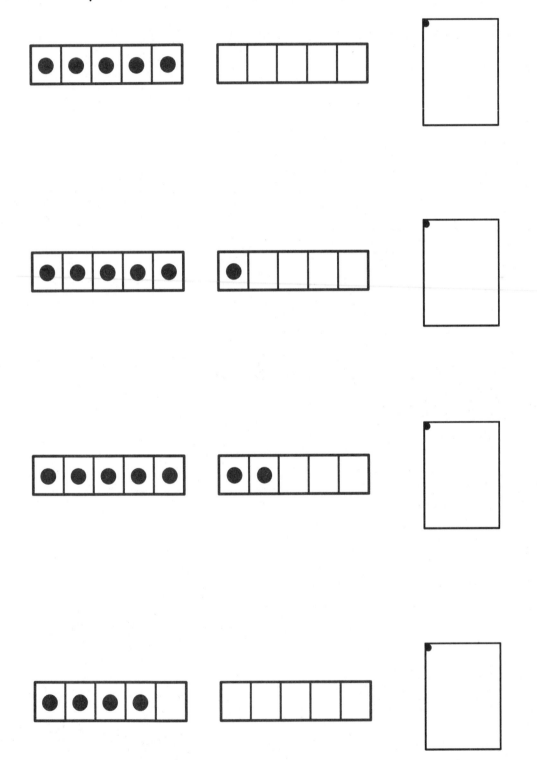

Lesson 21:   Compare counts of 8 in linear and array configurations. Match with numeral 8.

©2016 Great Minds. eureka-math.org
GK-M1-SE-B1-1.3.1-01.2016

Name _____   Date _____

Write 8.

8

Color 8 happy faces.

Circle a different group of 8 happy faces.

_____

numeral formation practice sheet 8

EUREKA
MATH™

Lesson 22:   Arrange and strategize to count 8 beans in circular (around a cup) and
scattered configurations. Write numeral 8. Find a path through the
scatter set, and compare paths with a partner.

105

©2016 Great Minds. eureka-math.org
GK-M1-SE-B1-1.3.1-01.2016

This page intentionally left blank

Name _____   Date _____

Draw a counting path with a line to show the order in which you counted. Write the total number. Circle a group of 5 in each set.

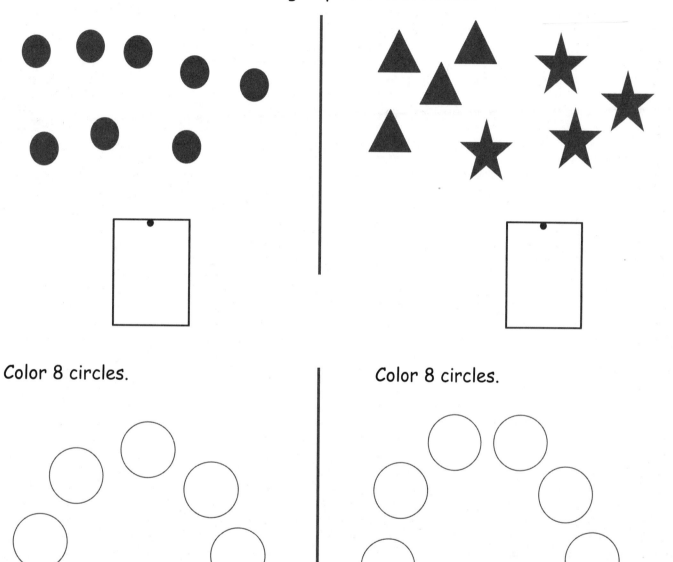

Color 8 circles.

Color 8 circles.

                Lesson 22:   Arrange and strategize to count 8 beans in circular (around a cup) and   107
EUREKA MATH™   scattered configurations. Write numeral 8. Find a path through the
                                                    scatter set, and compare paths with a partner.

©2016 Great Minds. eureka-math.org
GK-M1-SE-B1-1.3.1-01.2016

This page intentionally left blank

Name _____   Date _____

Draw 8 beads around the circle.

Color 8. Draw a line to show your counting path.

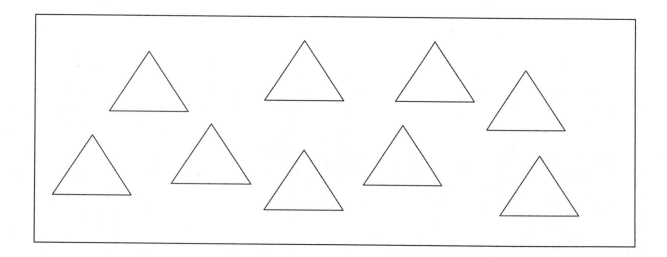

Lesson 22:   Arrange and strategize to count 8 beans in circular (around a cup) and
scattered configurations. Write numeral 8. Find a path through the
scatter set, and compare paths with a partner.

©2016 Great Minds. eureka-math.org
GK-M1-SE-B1-1.3.1-01.2016

109

Count how many. Write the number in the writing rectangle.

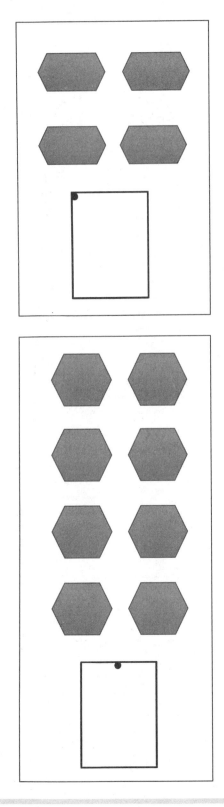

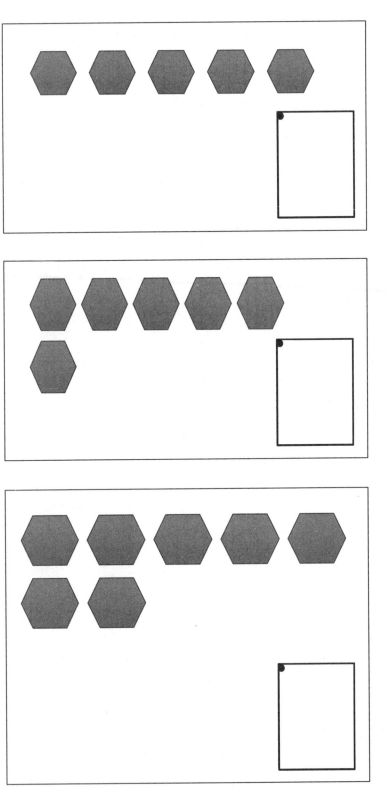

Lesson 22: Arrange and strategize to count 8 beans in circular (around a cup) and scattered configurations. Write numeral 8. Find a path through the scatter set, and compare paths with a partner.

Name _____   Date _____

Write 8.

[8]   [ ]   [ ]   _____   _____

[8]   [ ]   [ ]   _____   _____

Color 8 happy faces.

Circle a different group of 8 happy faces.

numeral formation practice sheet 8

Lesson 22: Arrange and strategize to count 8 beans in circular (around a cup) and
scattered configurations. Write numeral 8. Find a path through the
scatter set, and compare paths with a partner.

105

©2016 Great Minds. eureka-math.org
GK-M1-SE-B1-1.3.1-01.2016

This page intentionally left blank

Name _____   Date _____

Count and circle how many. Color 5.

7   8   9

7   8   9

Draw 4 circles to the right.
Count all the circles.

Draw 4 circles below. Count all
the circles.

Lesson 23:   Organize and count 9 varied geometric objects in linear and array
(3 threes) configurations. Place objects on 5-group mat. Match with
numeral 9.

111

©2016 Great Minds. eureka-math.org
GK-M1-SE-B1-1.3.1-01.2016

Color 3. Count and circle how many.

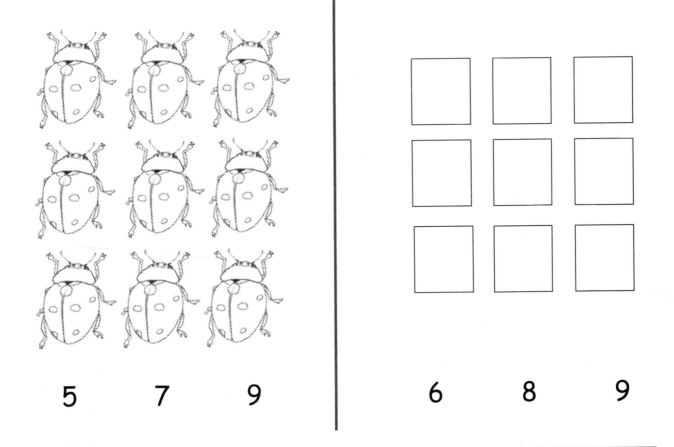

5   7   9   6   8   9

Color 3. Draw 2 circles to finish the last row. Count how many.

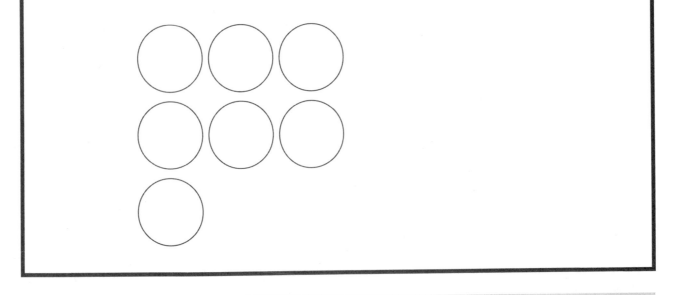

Lesson 23:   Organize and count 9 varied geometric objects in linear and array
(3 threes) configurations. Place objects on 5-group mat. Match with
numeral 9.

Name _____   Date _____

Color 9 shapes.

Color 9 shapes.

Draw 9 circles.

Draw 9 shapes.

EUREKA MATH

Lesson 23: Organize and count 9 varied geometric objects in linear and array
(3 threes) configurations. Place objects on 5-group mat. Match with
numeral 9.

113

This page intentionally left blank

Name _____   Date _____

Write 9.

9 | | | ___ ___

9 | | | ___ ___

Color 9 happy faces.

Circle a different group of 9 happy faces.

numeral formation practice sheet 9

EUREKA MATH

Lesson 24:   Strategize to count 9 objects in circular (around a paper plate) and
scattered configurations printed on paper. Write numeral 9.
Represent a path through the scatter count with each object.

115

This page intentionally left blank

Name _____   Date _____

Draw a counting path with a line to show the order in which you counted. Write the total number. Circle a group of 5 in each set.

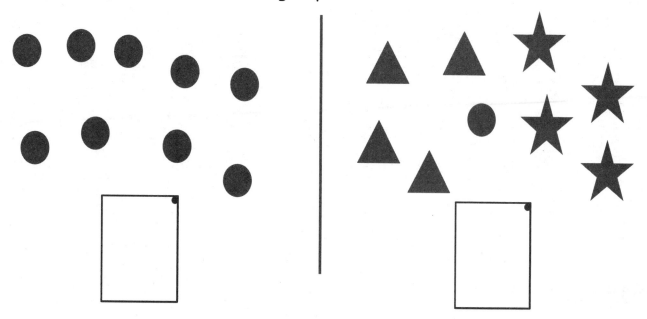

Count the stars and objects. Write the total number of objects in the boxes.

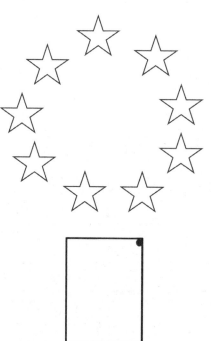

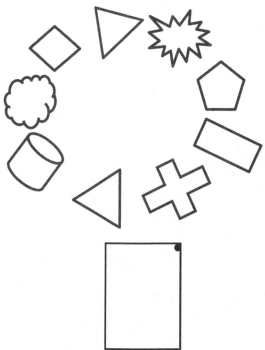

Lesson 24:   Strategize to count 9 objects in circular (around a paper plate) and
scattered configurations printed on paper. Write numeral 9.
Represent a path through the scatter count with each object.   117

©2016 Great Minds. eureka-math.org
GK-M1-SE-B1-1.3.1-01.2016

Count the dots.
Write the number.

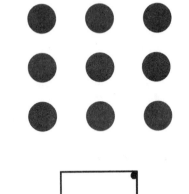

Count the dots. Write the number.
Circle a group of 5.

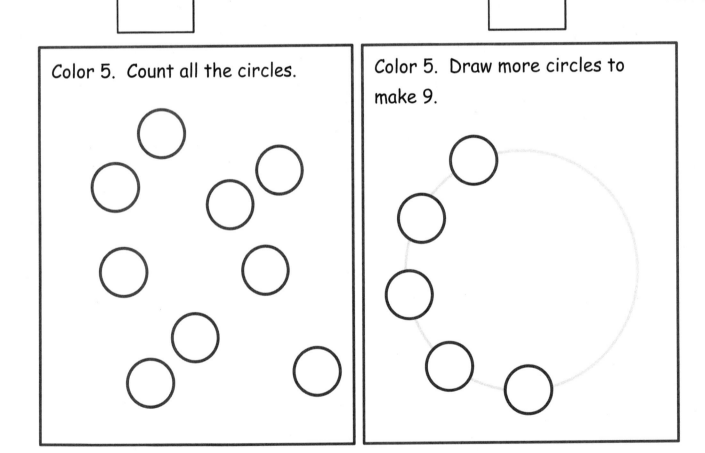

Color 5. Count all the circles.

Color 5. Draw more circles to make 9.

Lesson 24:   Strategize to count 9 objects in circular (around a paper plate) and
scattered configurations printed on paper. Write numeral 9.
Represent a path through the scatter count with each object.

EUREKA
MATH™

Name _____   Date _____

Color 9 circles.

Color 9 circles.

Draw 9 beads on the bracelet.

Count. Write the number in the box.

Lesson 24:   Strategize to count 9 objects in circular (around a paper plate) and scattered configurations printed on paper. Write numeral 9. Represent a path through the scatter count with each object.

119

©2016 Great Minds. eureka-math.org
GK-M1-SE-B1-1.3.1-01.2016

This page intentionally left blank

Name _____   Date _____

Count and circle how many. Color 5.

8   9   10

8   9   10

Color 5 circles. Draw 5 circles to the right. Count all the circles.	Color 5 circles. Draw 5 circles below. Count all the circles.

Lesson 25:   Count 10 objects in linear and array configurations (2 fives). Match with numeral 10. Place on the 5-group mat. Dialogue about 9 and 10. Write numeral 10.

©2016 Great Minds. eureka-math.org
GK-M1-SE-B1-1.3.1-01.2016

Count and circle how many. Color 5.

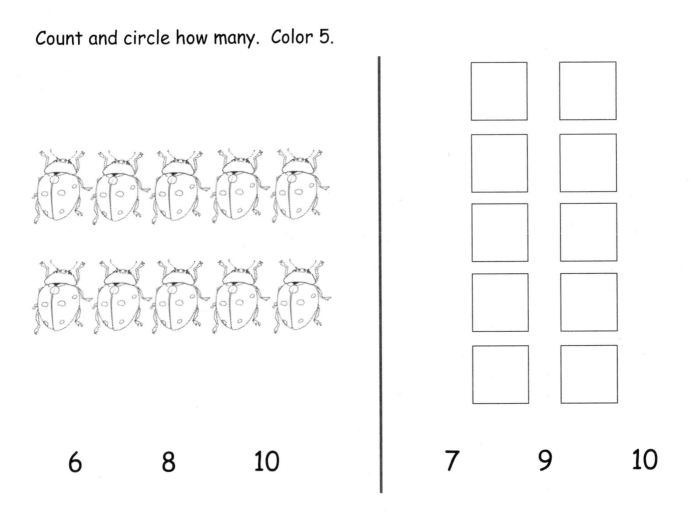

6   8   10   7   9   10

Color 5 circles. Draw 4 circles to finish the row.

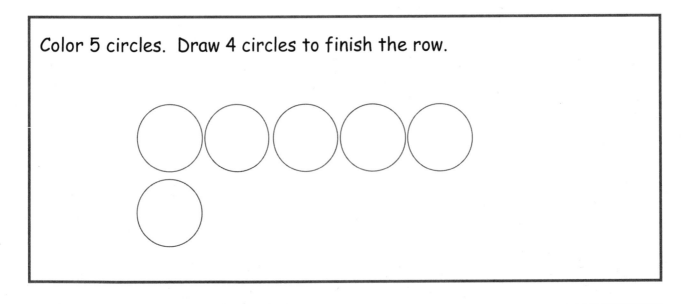

Lesson 25:   Count 10 objects in linear and array configurations (2 fives). Match
with numeral 10. Place on the 5-group mat. Dialogue about
9 and 10. Write numeral 10.

Name _____   Date _____

Color 9 squares. Color 1 more square a different color.

Draw 10 circles in a line. Color 5 circles red. Color 5 circles blue.

Draw 5 circles under the row of circles. Color 5 circles red. Color 5 circles blue.

Lesson 25:   Count 10 objects in linear and array configurations (2 fives). Match
with numeral 10. Place on the 5-group mat. Dialogue about
9 and 10. Write numeral 10.

123

©2016 Great Minds. eureka-math.org
GK-M1-SE-B1-1.3.1-01.2016

This page intentionally left blank

Name _____   Date _____

Write 10.

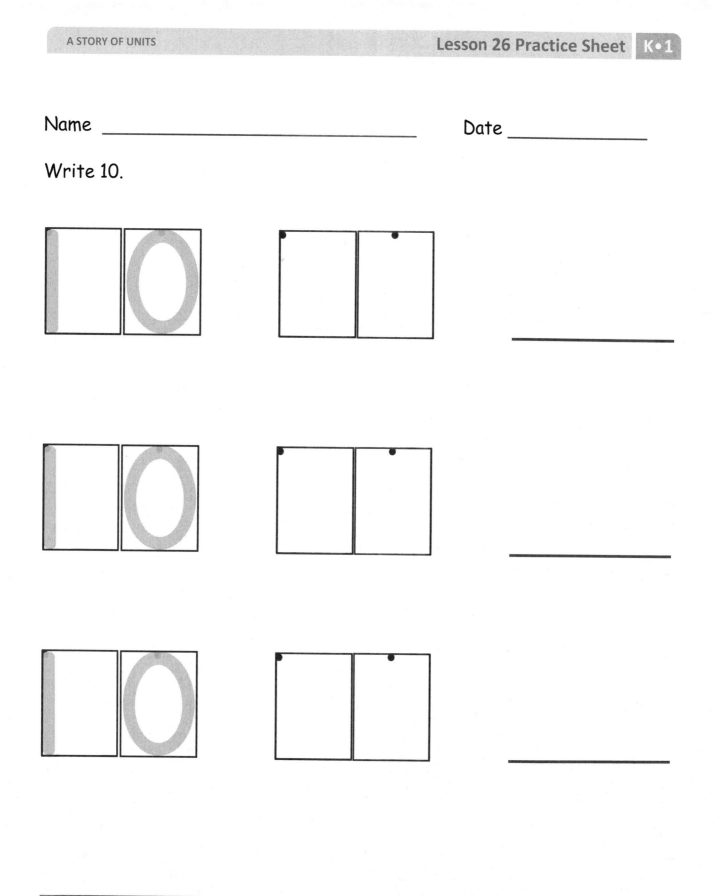

numeral formation practice sheet 10

Lesson 26: Count 10 objects in linear and array configurations (2 fives). Match with numeral 10. Place on the 5-group mat. Dialogue about 9 and 10. Write numeral 10.

125

©2016 Great Minds. eureka-math.org
GK-M1-SE-B1-1.3.1-01.2016

This page intentionally left blank

Name _____   Date _____

Draw 10 circles in a row. Color the first 5 yellow and the second 5 blue.
Write how many circles in the boxes.

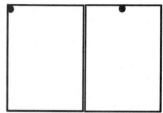

Draw 5 circles in the gray part. Draw 5 circles in the white part.
Write how many circles in the boxes.

Draw two towers of 5 next to each other. Color 1 tower red and the other tower orange.	Draw a row of 5 cubes. Draw another row of 5. Count.

Draw a picture of your bracelet on the back of your paper.

EUREKA
MATH™

Lesson 26:   Count 10 objects in linear and array configurations (2 fives). Match
with numeral 10. Place on the 5-group mat. Dialogue about 9 and 10.
Write numeral 10.

This page intentionally left blank

Name _____   Date _____

Draw 5 triangles in a row. Draw another 5 triangles in a row under them.

How many triangles did you draw? Write the number.

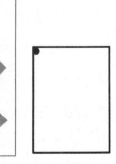

Write how many.   Write how many.

   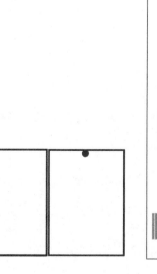

Lesson 26:   Count 10 objects in linear and array configurations (2 fives). Match   **129**
with numeral 10. Place on the 5-group mat. Dialogue about 9 and 10.
Write numeral 10.

©2016 Great Minds. eureka-math.org
GK-M1-SE-B1-1.3.1-01.2016

This page intentionally left blank

Name _____   Date _____

Count the shapes, and write how many. Color the shape you counted first.

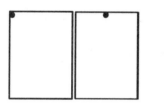

Draw 10 things. Color 5 of them. Color 5 things a different color.

Draw 10 circles. Color 5 circles. Color 5 circles a different color.

Color 10 apples. Draw a path to connect the apples starting at 1.

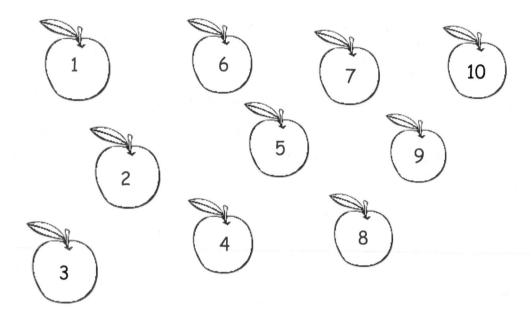

Color 10 apples. Count and draw a path to connect the apples.

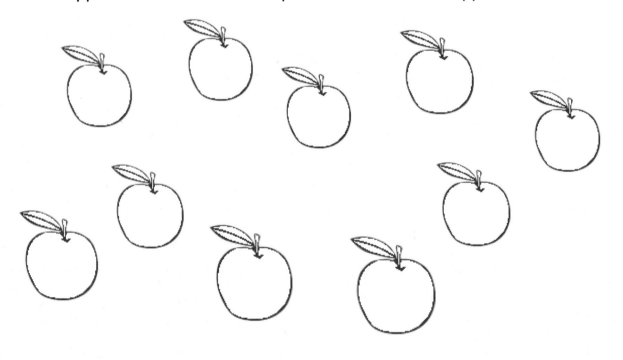

Name _____   Date _____

Draw more ☁ to make 10.

Draw more 🙂 to make 10.

This page intentionally left blank

Name _____   Date _____

Listen to my stories. Color the pictures to show what is happening. Write how many in the box.

Bobby picked 4 red flowers. Then, he picked 2 purple flowers. How many flowers did Bobby pick?

Janet went to the donut store. She bought 6 chocolate donuts and 3 strawberry donuts. How many donuts did she buy?

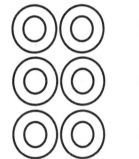

Some children were sitting in a circle. 4 of them were wearing green shirts. The rest were wearing yellow shirts. How many children were in the circle?

Jerry spilled his bag of marbles. Circle the group of grey marbles. Circle the group of black marbles. How many marbles were spilled?

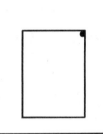

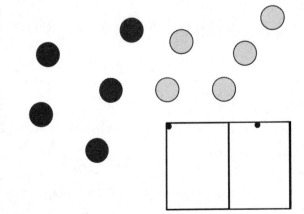

Lesson 28:   Act out *result unknown* story problems without equations.

135

©2016 Great Minds. eureka-math.org
GK-M1-SE-B1-1.3.1-01.2016

Make up a story about the bears. Color the bears to match the story.
Tell your story to a friend.

Make up a new story. Draw a picture to go with your story. Tell your
story to a friend.

©2016 Great Minds. eureka-math.org
GK-M1-SE-B1-1.3.1-01.2016

Name _____   Date _____

Make up a story about 10 things in your house. Draw a picture to go with your story. Be ready to share your story at school.

This page intentionally left blank

Name _____   Date _____

Count the dots. Write how many. Draw the same number of dots below, but going up and down instead of across. The number 4 has been done for you.

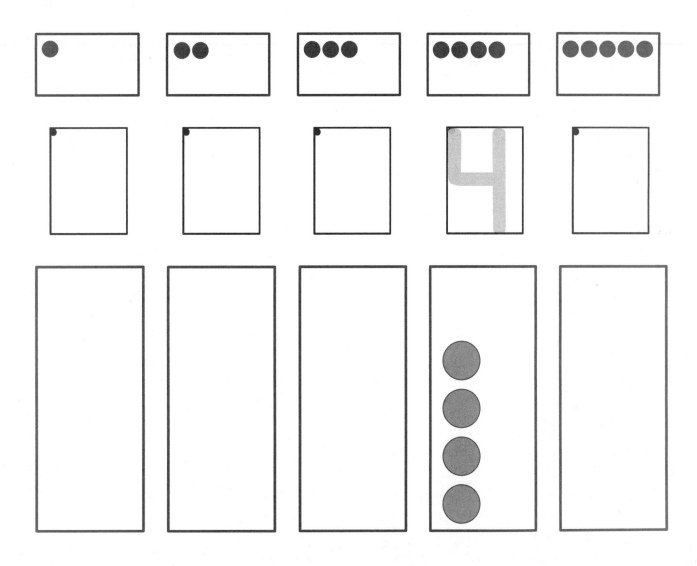

Lesson 29: Order and match numeral and dot cards from 1 to 10. State 1 more than a given number.

139

©2016 Great Minds. eureka-math.org
GK-M1-SE-B1-1.3.1-01.2016

Count the objects. Draw 1 more object. Count and write how many.

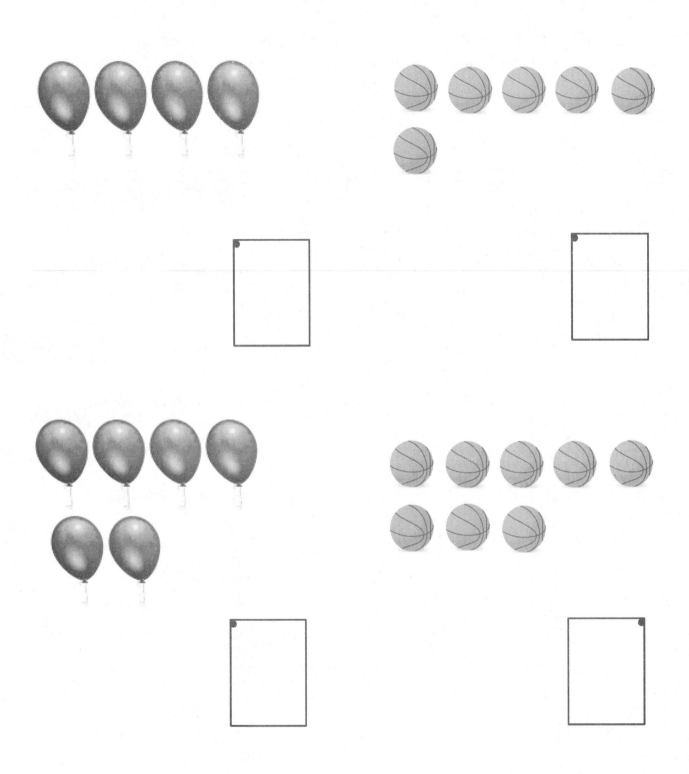

Order and match numeral and dot cards from 1 to 10. State 1 more than a given number.

Name _____   Date _____

Count the dots. Write how many. Draw the same number of dots below, but going up and down instead of across. The number 6 has been done for you.

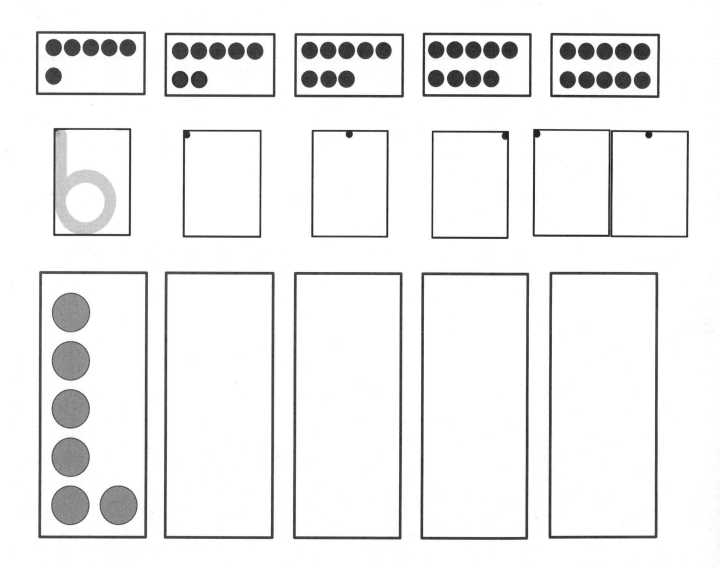

Lesson 29:   Order and match numeral and dot cards from 1 to 10. State 1 more than a given number.

141

©2016 Great Minds. eureka-math.org
GK-M1-SE-B1-1.3.1-01.2016

This page intentionally left blank

Make your own 5-group cards! Cut the cards out on the dotted lines. On one side, write the numbers from 1 to 10. On the other side, show the 5-group dot picture that goes with the number.

EUREKA
MATH™

Lesson 29:   Order and match numeral and dot cards from 1 to 10. State 1 more
                  than a given number.

143

©2016 Great Minds. eureka-math.org
GK-M1-SE-B1-1.3.1-01.2016

This page intentionally left blank

Name _____   Date _____

Count and color the white squares red. Count all the cubes in each step.
Write the missing numbers below each step.

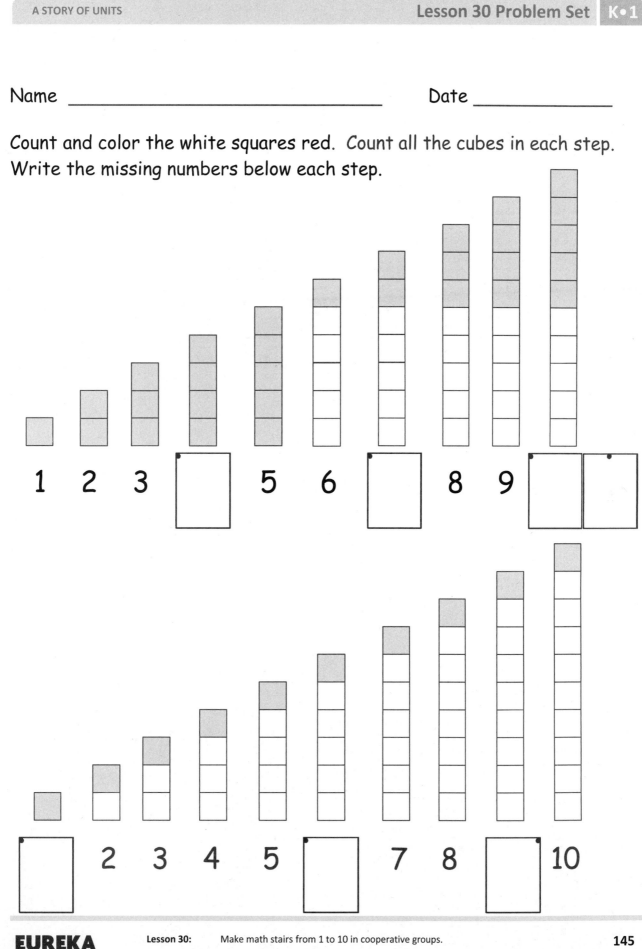

Lesson 30:   Make math stairs from 1 to 10 in cooperative groups.

145

EUREKA
MATH™

This page intentionally left blank

Name _____   Date _____

Draw the missing stairs. Write the numbers below each step.

Ask someone to help you write about what you think baby bear will do now that you have helped him to get home. Use the back of this paper.

Draw 1 more cube on each stair so the cubes match the number. Say as you draw, "1. One more is two. 2. One more is three."

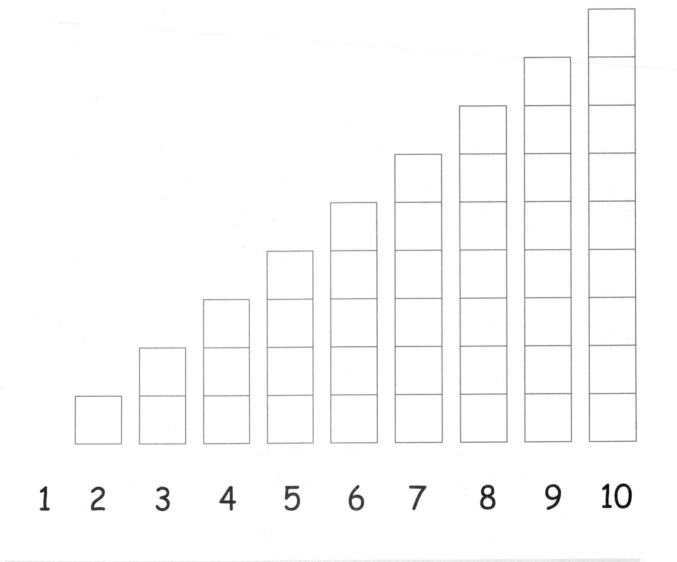

1   2   3   4   5   6   7   8   9   10

EUREKA MATH™

©2016 Great Minds. eureka-math.org
GK-M1-SE-B1-1.3.1-01.2016

Name _____ Date _____

Color and count the empty circles. Count the gray circles. Write how many gray circles in the box.

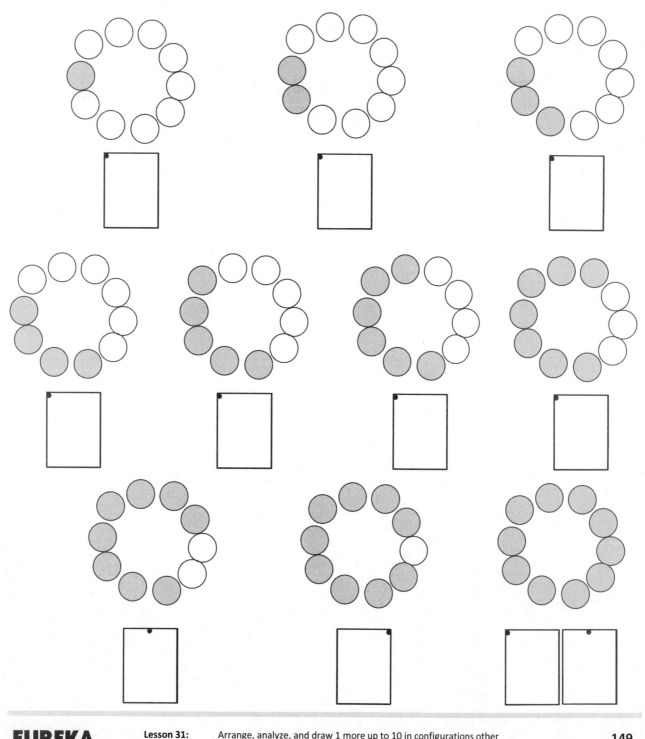

Lesson 31:   Arrange, analyze, and draw 1 more up to 10 in configurations other
             than towers.

149

©2016 Great Minds. eureka-math.org
GK-M1-SE-B1-1.3.1-01.2016

Draw 1 more circle and count all the circles. Write how many.

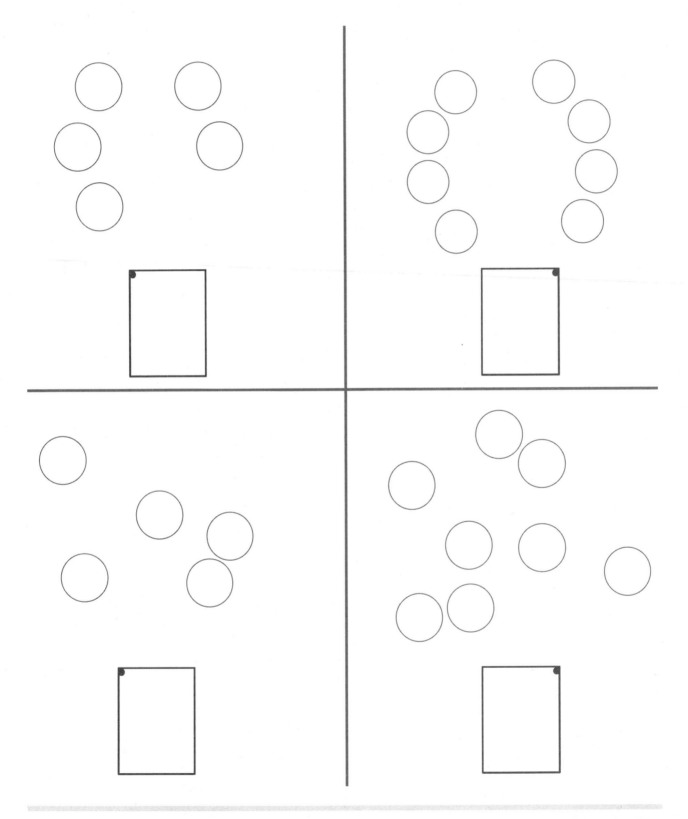

Arrange, analyze, and draw 1 more up to 10 in configurations other than towers.

Name _____   Date _____

Draw one more square. Color all the squares, and write how many.

Draw one more cloud. Color all the clouds, and write how many.

Lesson 31:   Arrange, analyze, and draw 1 more up to 10 in configurations other than towers.

151

©2016 Great Minds. eureka-math.org
GK-M1-SE-B1-1.3.1-01.2016

This page intentionally left blank

Name _____   Date _____

Draw and write the number of the missing steps.

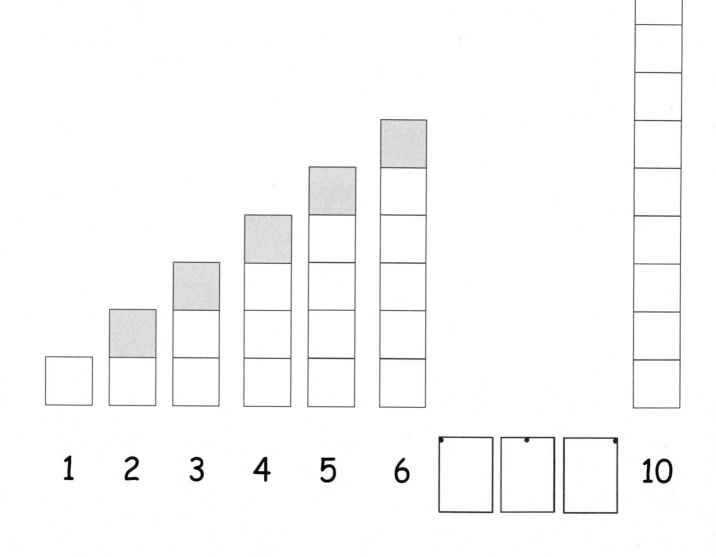

Lesson 32:   Arrange, analyze, and draw sequences of quantities of 1 more, beginning with numbers other than 1.

153

©2016 Great Minds. eureka-math.org
GK-M1-SE-B1-1.3.1-01.2016

Write the missing number. Draw objects to show the numbers.

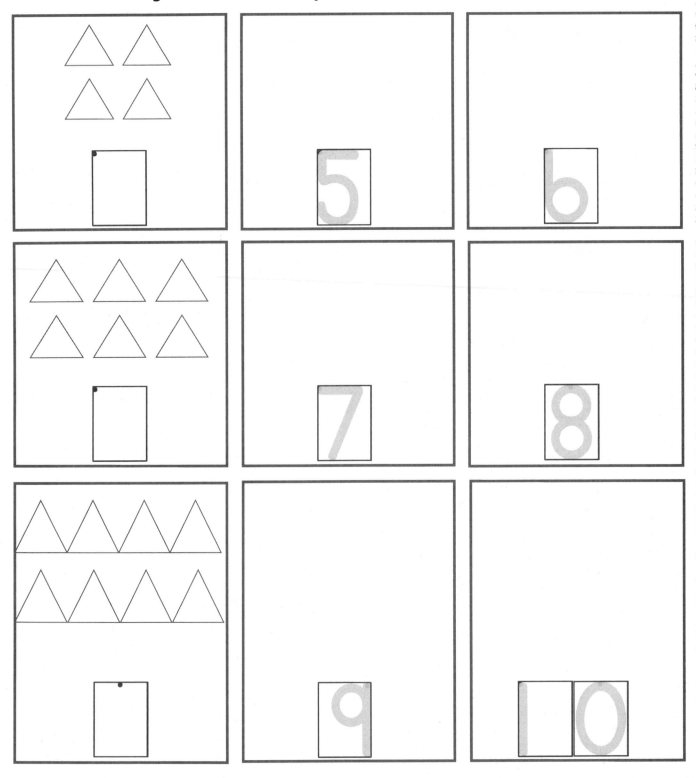

Lesson 32:   Arrange, analyze, and draw sequences of quantities of 1 more, beginning with numbers other than 1.

Name _____   Date _____

Write the missing numbers.

☐ ,2, ☐ , ☐ , ☐ , ☐ ,6,7, ☐ , ☐ ,10

Draw X's or O's to show 1 more.

XX	**X**		**XXXX** **X**
OOOO **O**	**OOO**	**OOOO** **OOO**	O O O O

Tell someone a story about "1 more...and then 1 more." Draw a picture about your story.

EUREKA
MATH™

Lesson 32:   Arrange, analyze, and draw sequences of quantities of 1 more,
             beginning with numbers other than 1.

155

©2016 Great Minds. eureka-math.org
GK-M1-SE-B1-1.3.1-01.2016

This page intentionally left blank

Name _____ Date _____

Count the dots. Write how many. Draw the same number of dots below, but go up. The number 6 is done for you.

Lesson 33: Order quantities from 10 to 1, and match numerals.

157

EUREKA
MATH™

©2016 Great Minds. eureka-math.org
GK-M1-SE-B1-1.3.1-01.2016

Count the dots. Write how many. Draw the same number of dots below, but go up. The number 4 is done for you.

Count the balloons. Cross out 1 balloon. Count and write how many balloons are left in the box.

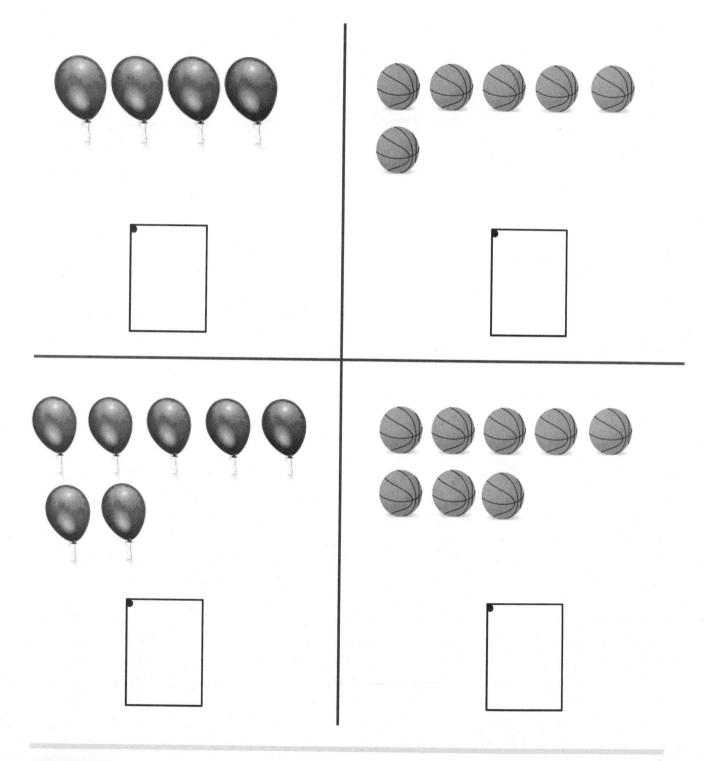

Lesson 33:   Order quantities from 10 to 1, and match numerals.

159

This page intentionally left blank

Make 5-group Cards

Cut the cards out on the dotted lines. On one side, write the numbers from 1-10. On the other side, show the 5-group dot picture that goes with the number. Mix up your cards, and practice putting them in order in the "1 less" way.

Lesson 33: Order quantities from 10 to 1, and match numerals.

161

©2016 Great Minds. eureka-math.org
GK-M1-SE-B1-1.3.1-01.2016

This page intentionally left blank

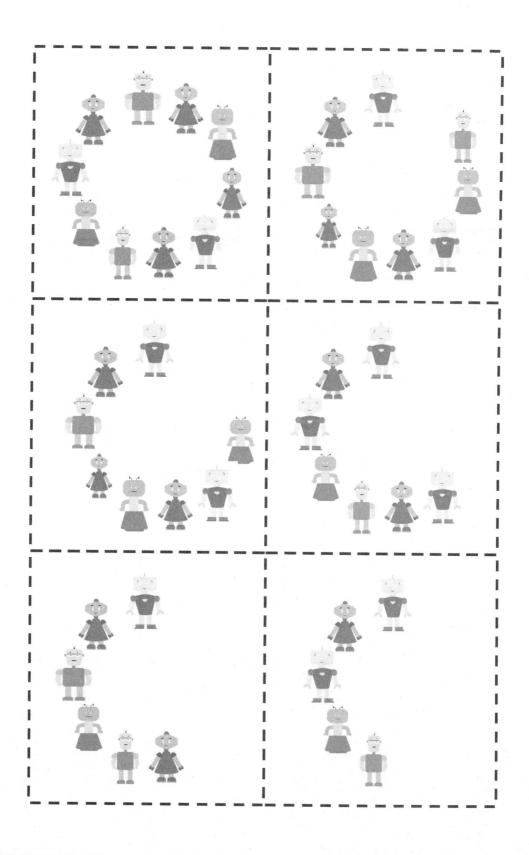

©2016 Great Minds. eureka-math.org
GK-M1-SE-B1-1.3.1-01.2016

This page intentionally left blank

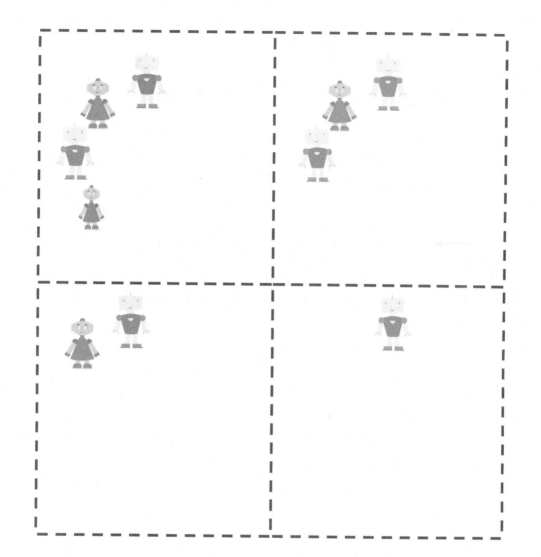

1	2	3	4	5
6	7	8	9	10

This page intentionally left blank

Name _____   Date _____

Count and write the number of apples. Color the group of apples that is 1 less.

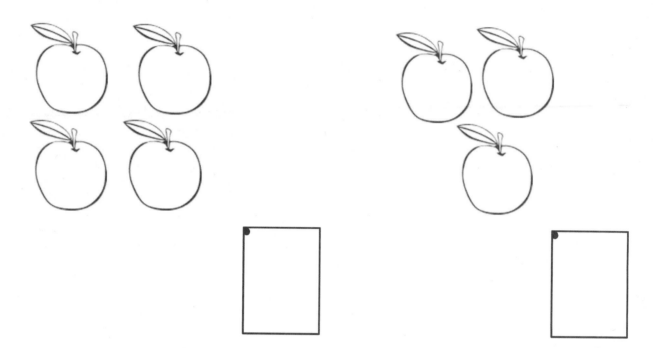

Count and write the number of hearts. Color the group of hearts that is 1 less.

EUREKA
MATH™

This page intentionally left blank

Name _____   Date _____

Count and write the number of objects. Draw and write the number of objects that is 1 less.

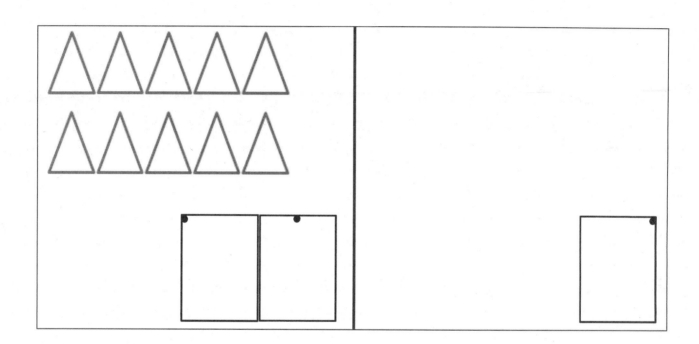

Lesson 34:   Count down from 10 to 1, and state 1 less than a given number.

169

This page intentionally left blank

Name _____   Date _____

Count all the squares in each tower, and write how many. What do you notice?

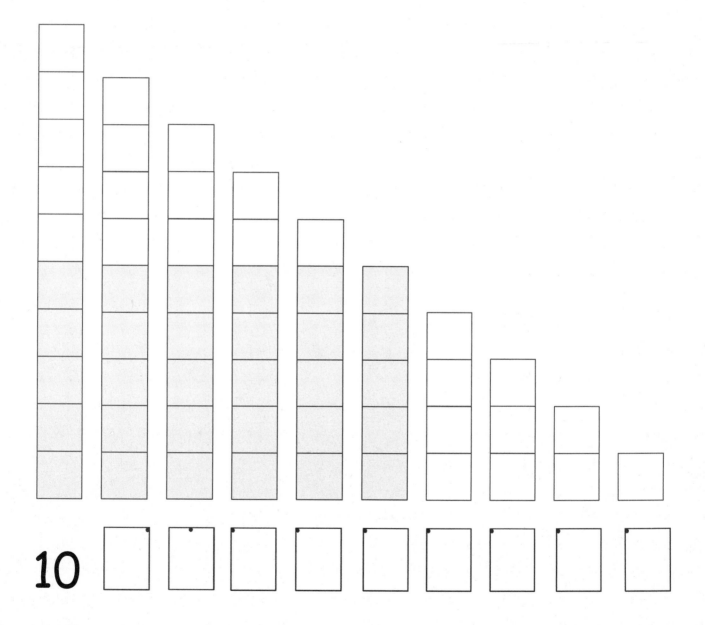

10

Lesson 35:   Arrange number towers in order from 10 to 1, and describe the
              pattern.

171

©2016 Great Minds. eureka-math.org
GK-M1-SE-B1-1.3.1-01.2016

Count the number of squares in each stair. Cross off the top square. Use your words to say, "10. One less is nine. 9. One less is eight." Write how many squares are in each stair after you cross off.

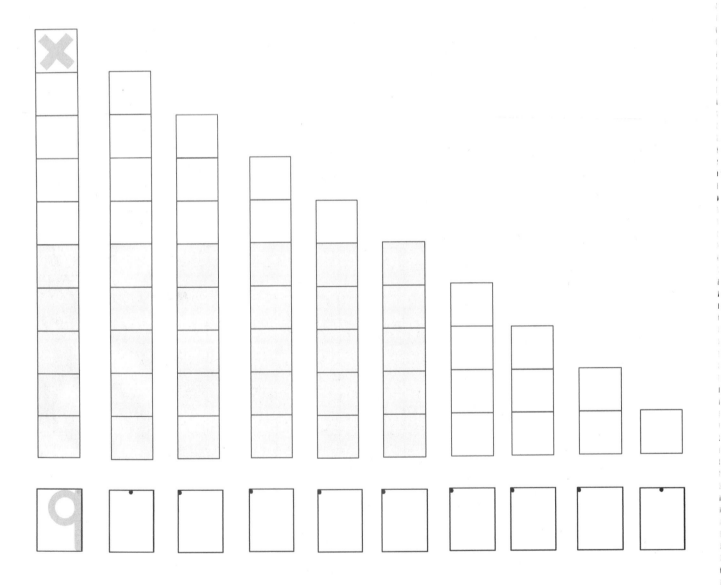

Lesson 35:   Arrange number towers in order from 10 to 1, and describe the pattern.

Name _____   Date _____

Count all the squares in each tower, and write how many. Share with someone what you notice!

10

EUREKA MATH™

Lesson 35: Arrange number towers in order from 10 to 1, and describe the pattern.

173

©2016 Great Minds. eureka-math.org
GK-M1-SE-B1-1.3.1-01.2016

This page intentionally left blank

Name _____   Date _____

Count all the objects. Write the number in the first box.

Count the objects that are white. Write that number in the second box.

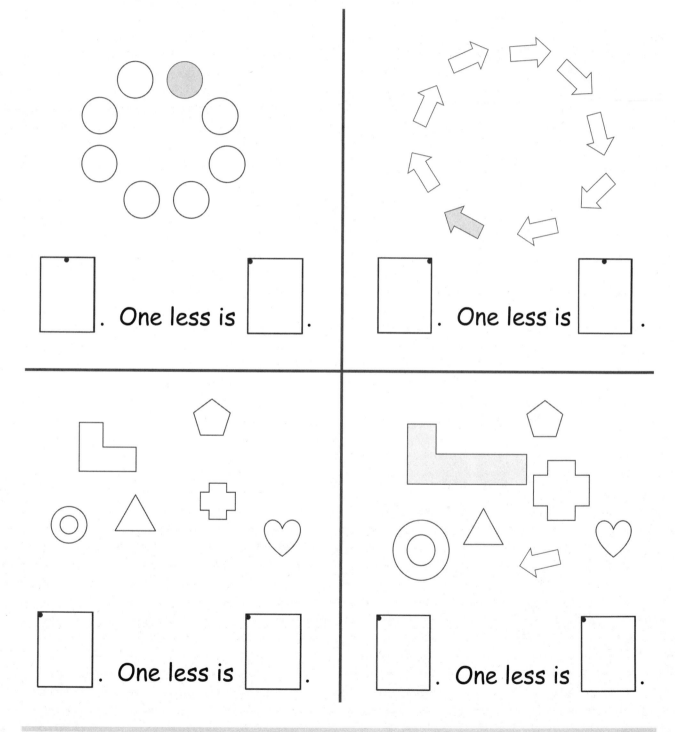

. One less is [ ] .

[ ] . One less is [ ] .

[ ] . One less is [ ] .

[ ] . One less is [ ] .

Lesson 36:   Arrange, analyze, and draw sequences of quantities that are 1 less configurations other than towers.

©2016 Great Minds. eureka-math.org
GK-M1-SE-B1-1.3.1-01.2016

175

Count and write how many.

Draw 1 less. Write how many.

Count and write how many.

Draw 1 less. Write how many.

Lesson 36:   Arrange, analyze, and draw sequences of quantities that are 1 less configurations other than towers.

Name _____   Date _____

Draw bracelets with the number of beads shown.

Write the missing number. Hint: The missing number is 1 less!

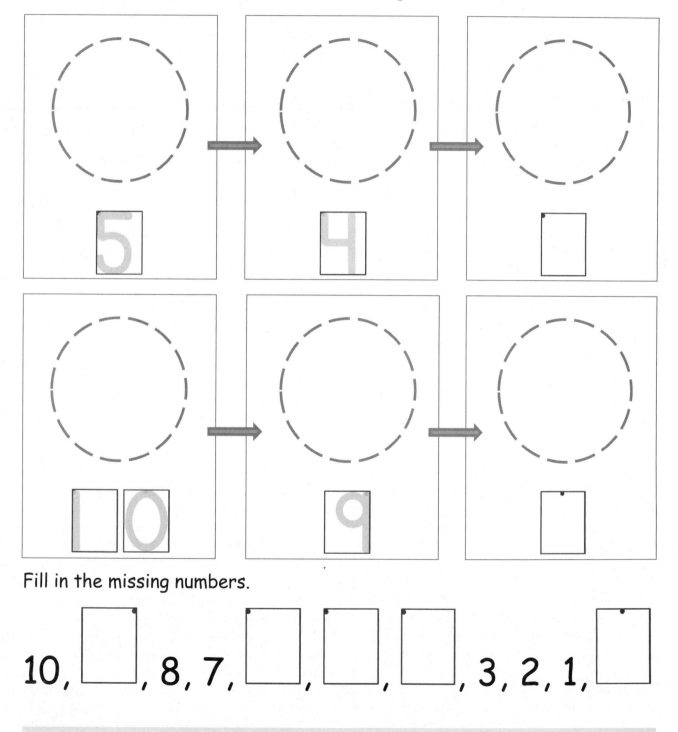

Fill in the missing numbers.

10, ☐, 8, 7, ☐, ☐, ☐, 3, 2, 1, ☐

Lesson 36: Arrange, analyze, and draw sequences of quantities that are 1 less configurations other than towers.

177

©2016 Great Minds. eureka-math.org
GK-M1-SE-B1-1.3.1-01.2016

This page intentionally left blank

Name _____ Date _____

Count how many are in each group. Write the number.

CHALLENGE: Circle the smaller group in each row.

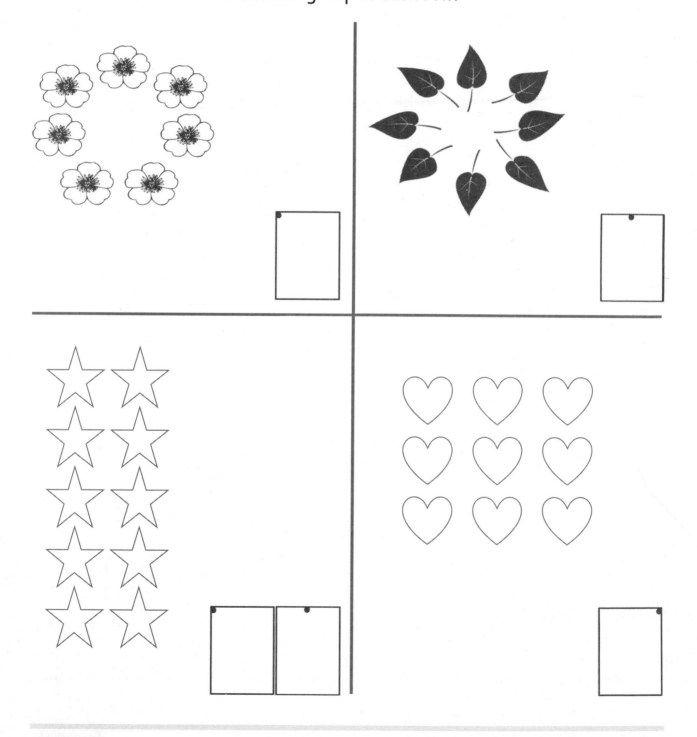

Draw some toys you enjoy.

How many?

Draw some healthy foods.

How many?

Lesson 37:   Culminating Task